SAS TRACKING HANDBOOK

Barry Davies, BEM

Skyhorse Publishing

Skyhorse Publishing books may be purchased in bulk at special discounts for sales promotion, corporate gifts, fund-raising, or educational purposes. Special editions can also be created to specifications. For details, contact the Special Sales Department, Skyhorse Publishing, 307 West 36th Street, 11th Floor, New York, NY 10018 or info@skyhorsepublishing.com.

Skyhorse® and Skyhorse Publishing® are registered trademarks of Skyhorse Publishing, Inc.®, a Delaware corporation.

Visit our website at www.skyhorsepublishing.com.

10 9 8 7 6 5 4 3 2 1

Library of Congress Cataloging-in-Publication Data is available on file.

ISBN: 978-1-62914-235-7

Printed in China

For my daughter Sarah,
in the hope that you will also become a writer.

TABLE OF CONTENTS

INTRODUCTION

In an age where humans rely more on technology and robotics, the potential of visual tracking seems to be an old hat with its true value totally misunderstood. There are so many misconceptions about tracking that the art has somehow lost its attraction, especially within the military. Visual tracking is seen as a skill associated with the Native American, San Bushmen, Iban, or Dyak trackers; an era in the past with no modern day significance. Today, electronic tracking, mobile phones, and drones take priority over the human eye.

Yet a few have kept the skill alive, most of whom are retired Special Forces personnel that have successfully practiced their tracking skills during military operations. In the 1950s through the end of the Vietnam War, military commanders used the skills of the tracker to gain vital intelligence on the enemy, locate their position, and thus enable search-and-destroy missions. This ability to locate, identify, pursue, and interpret those signs, as well as form reasonably accurate conclusions based on the evidence left by the quarry was invaluable. Now it is all but gone, replaced by drones and other high-tech battlefield surveillance.

As modern warfare turns from raging tank battles to more isolated counter-terrorism, the need for visual trackers once more becomes a relevant military skill. No terrorist can move across any terrain without leaving some type of evidence. Gathering this evidence may well mean email or phone interception or the use of high flying drones; but in the jungles and mountains where technology is sparse; there is still a place for the visual tracker. In such areas, signs left by the quarry can reveal much about the enemy.

Historically, visual human tracking has been used by many military and law enforcement agencies in other countries around the world with a great deal of success. The ability to employ visual trackers to locate and interdict a subject attempting to elude their pursuers, gather information for intelligence purposes, or help rescue lost individuals and groups is essential. While it may not be noticeable, visual tracking in one form or another is still widespread with border police around the world, constantly searching millions of miles on a daily basis in an effort to stop illegal crossings.

The tracker's art is simple: from the signs left behind they will follow a trail and, in doing so, build a picture and ask—*How many persons am I following? How are they equipped? What is their state of morale? Do they know they are being followed? Where are they going?* To answer these questions, the tracker uses available indicators; that is, signs that tell if any action has occurred and at a specific time and place. By comparing indicators, the tracker obtains all the answers to his questions.

Tracking started with man's need for food; to have the ability to understand what they were following and what would be the rewards if they were successful. In tracking terms, little has changed: we still track game for sport and food, but man has found other uses for tracking (border polity, military tracking terrorists, etc). Over the years, tracking became both a civil and military skill, sadly one that only raises its head when needed. Military trackers were successfully used in Malayan, Kenya, Cyprus, Borneo, Vietnam, and Rhodesia to name just a few. Military trackers in particular proved very successful because they were able to pass back valuable information such as strength, speed, and other pertinent information required to successfully bring the fight to their foe.

In short, visual tracking is the ability to follow and locate a man or animal by observing its path by the signs they have left behind. These signs are classed as top or bottom sign, temporary or permanent; but sign can also mean direction, display habits, leave scent or smell, and many other factors that will help identify the path taken.

In many tracking units, man's best friend is the tracker dog. Humans and dogs work well together and often speed up the tracking process, as dogs track by smell and not sign. Yet more and more today, man relies on technology and most of the world borders and war zones are inundated with surveillance drones. These *eyes in the sky* move very quickly and are capable of detecting a moving target from a distance. Powerful camera gimbals support both day and night observation, seeing clearly into the darkness displaying the heat signature on a computer terminal hundreds of miles away. We slip our credit card casually into an ATM while on holiday in Asia and immediately people have your rough location. Finally, try walking undetected in the UK with its 4.2 million closed circuit TV cameras, 265,000 of which are in London alone; that's one for every fourteen people!

As our world matures, so does the amount of surveillance. Believe me, we are all being tracked.

Chapter 1

WHAT IS TRACKING? HOW IS IT USED?

The success or failure of the modern tracker is dependent on their personal skills. Training is vital in learning tracking skills, and continuous exercise is the best way to interpret sign. These skills are rarely found, but remain hidden deep within all of us. The British SAS tracked down terrorists in the jungles of Malaysia and Indonesia, while the Selous Scouts used their skills in the bush of Rhodesia. Today, there are many specialist military and police units operating around the world each practicing and refining their tracking skills.

Tracking is a skill that has been passed down from generation to generation and to some degree, it takes place every day. Most people are completely unaware of the amount of tracking that takes place around the world. Apart from hunting skills, the main use of tracking today is to track down humans. This may mean protecting our national borders against illegal immigrants, looking for terrorist training camps in the desert or jungle, or trying to locate climbers lost on an inhospitable mountain. No matter the reason, the basics of human tracking remain very similar to as they did thousands of years ago.

Then there is modern-day tracking, which uses sophisticated electronic devices to aid in finding the whereabouts of an individual no matter where they are in the world. While this is still referred to as *tracking*, it is a completely different science. Today thousands of individuals are "tagged" for tracking because they have committed some form of criminal act. Governments attached small and in some cases minuscule tracking devices onto a car or even on the individual, so that their whereabouts can be tracked without even leaving the office.

Most people relate human and animal tracking to the forests and wilderness, or some war zone in a far-off land. While this is basically true in the real sense of tracking, but we also track

Kuna Indian hunter from Colombia, skilled at tracking and familiar with what was in his own backyard.

people in urban areas and the cities of the world. The normal signs left behind by man and animal do exist, and if they do, they are soiled by the multitude of people living and working in a close community, such as a town or city. Yet in the cities, tracking is possibly more prevalent than in the wilderness.

Some criminals are fitted with electronic bracelets on their ankles so that their whereabouts can be tracked at all times. CCTV cameras monitor most city centers and major highways, and many vehicles are fitted with GPS trackers in case they are stolen. For those that travel, our passports become ever more sophisticated as electronic chips are introduced, allowing governments to identify our international travel. More and more we hear of drones flying silently high above us, transmitting live images to some control room thousands of miles away. This type of tracking is on the rise and will continue to dominate our lives in one form or another.

We have only to look at the current world news to see the extent of surveillance being undertaken by various agencies. Presently, the United States of America is under severe pressure from Europe to explain why it is monitoring the conversations of the German Chancellor Angela Merkel and other world leaders. They claim it is to protect the citizens of the United States and its allies, though few believe that is the real reason.

Tracking and listening in to mobile phone conversations has been common practice for many years and it makes no difference who you are: royalty, the President, or someone who is a danger to society. Those that track the location of your mobile do so by triangulation of the phone masts your mobile is close to—or in some cases, accessing the GPS location in your phone. They sit in one of the many offices run by the NSA or their little sister GCHQ in UK, or there sub-monitoring office in Oswestry, Shropshire. Mobile phones have become

GCHQ has many listening sites all over the United Kingdom and has extremely close ties with America's NSA.

very sophisticated, as have the app's designed to run on them—and we are all hungry to have the best and latest of both—but in doing so you run the risk of being tracked.

I always tell my students to think about the things they do every day and explain how we all track subconsciously. For example, if you drive to a supermarket parking lot, you drive on the road and down a market lane to select an empty spot to park your car. You leave your car and head for the store entrance which is clearly obvious—nevertheless, you are actually tracking. Why? Because you would not suddenly turn off the highway and drive over a water ditch and through a fence to gain access to the supermarket parking lot, although technically it would be a shorter route, it was full of obstacles. Once in the lot, you do not deliberately plough over other vehicles to find a parking space, and, to top it all, there is only one way into the supermarket unless you can walk through walls. Tracking in the wild is little different, as your target (unless deliberately trying to avoid being tracked) will do the same. They will take the most logical route, the obvious opening and the clearest, easiest course open to them. It is only when your quarry suspects they are being followed that they may choose to do things that will throw you off the scent.

MILITARY

The military has used human visual tracking skills for many years. In 1755, a New Hampshire captain by the name of Robert Rodgers recruited retired local soldiers and formed them into a militia which became known as Rogers' Rangers. Most of the recruits were local frontiersmen who had a good knowledge of the local area, added to which they were all skilled trackers. They were employed by the British against the French and patrolled from outpost to outpost, looking for signs of enemy activity; many times they would follow a trail and engage the Indians supporting the French in combat. Their excellence at long-range patrolling laid down many of the rules modern soldiers refer to as Standard Operating Procedure (SOPs). When the American Revolution started, many of Rogers' Rangers joined up to fight against the British under the leadership of John Stark and eventually became part of the Continental Army.

Modern visual tracking came to the fore in the late 1960s in the jungles of Borneo. This means of locating the enemy lasted well into the Vietnam War and then started to decline. Today, few military units actually practice the art of human tracking and sadly it is a dying skill. Those that do continue with the skill tend to concentrate on either thick bush or jungle tracking in areas where human presence is fairly scarce.

Selous Scouts— the frontrunners of modern-day visual tracking, with a proven track record against an enemy in extremely hostile terrain.

In many military operations, soldiers are alert for signs of enemy activity. These signs help the soldier become aware of the enemy's presence and therefore give them time to react. The skill of tracking also allows the soldier to follow the enemy after contact has been broken and take the fight to the enemy's camp. During the Rhodesian War, the Selous Scouts were very skillful at doing this.

When it comes to military tracking, the Selous Scouts (1 SAS Regiment) were one of the most successful units ever. They were mainly engaged in a wide range of operations, from what was known as 'fire-force' actions on open battlefields to clandestine missions, deep within enemy territory. This ragged-looking force actually consisted of highly professional soldiers who showed exceptional courage against a bitter and unforgiving enemy.

At the height of the War in 1976, the Selous Scouts numbered some 700 men. They worked in small units of four to six men who would parachute or heli-hop into the bush in hot pursuit of ZIPRA and ZANLA guerrillas. The Selous Scouts were lightly equipped, carrying mostly ammunition and water that enabled them to quickly track and close in on the fleeing guerrillas. Once spotted, the Scouts would call for soldiers of C Squadron SAS (Rhodesian) to parachute forward towards the guerrillas, in order to cut them off. The Selous Scouts methods were so effective that they accounted for killing more guerrillas than the rest of the Rhodesian Army put together. Along with the Rhodesian SAS, the Selous Scouts were disbanded in 1980 when Prime Minister Ian Smith handed over to Robert Mugabe's government and Rhodesia became Zimbabwe. Most of the Selous Scouts made their way into the South African Army.

Author's Note: Captain Dave Dobson was one of the most outstanding officers of the Rhodesian SAS, having taken part in almost all of the actions during the conflict. In March of 1977, the Rhodesian military decided to deal with a strong ZANLA (Zimbabwe African National Liberation Army) garrison, numbering some 100 terrorists, based one kilometer south of the town of Chioco in the northern Mozambique province of Tete. A and B Troops of C Squadron Rhodesian SAS, numbering twenty-two all ranks in total, were given the task of carrying out a raid on the terrorist camp.

At dusk on the March 22, the assault force, under Captain Dave Dodson, was inserted by helicopter in two lifts from a forward base at Mtoko. Having been dropped approximately seventeen kilometers west of their objectives, Captain Dodson and his men moved off to an LUP in some thick undergrowth about a kilometer's distance away where they lay up until the moon appeared. Marching throughout the rest of the night, they halted just before dawn and lay up until dusk on the following evening.

At 2300 hours on March 23, the force moved up to its objective. In the distance, coming from the direction of Chioco, they could hear music and singing, which indicated that a major celebration was being held in the town. Captain Dodson sited a three-man 60mm light mortar team, whose task was to shell Chioco and its police station so as to prevent any attempt at reinforcement from the town during the attack on the camp. The camp itself consisted of a parade ground surrounded by barrack blocks on the three sides with the guard room being positioned nearest the town on the northern side. Between the buildings and the wire perimeter laid a series of defensive bunkers.

The groups moved to within 500 meters of the camp, setting up a mortar position. As the four assault groups moved past the mortar position, they dropped off their packs and quantities of mortar bombs before moving up to a three-strand perimeter wire fence and taking up their positions. Just before first light, Captain Dodson's and Sergeant Iain Bowen's groups slid under the wire and positioned ten Claymore mines along the back walls of two barrack blocks on the western side of the parade ground. At the same time, Corporal Nick Breytenbach was setting eight more Claymores on the northern corner of the camp.

The attack was launched at first light. Corporal Breytenbach's Claymores were initiated first, followed a split second later by those of Sergeant Bowen. At the same time, Corporal Frank Booth tossed two fragmentation grenades into a bunker holding a number of terrorists. The four assault groups then commenced their advance through the camp, firing at everything that moved. Having cleared the barrack buildings, they turned their attention to the terrorists' defensive positions, which comprised a network of trenches leading from inside the camp to outside the wire fence. The assault group threw grenades and 'bunker bombs,' 1kg explosive charges fitted with four-second fuses, into the trenches as the terrorists attempted to escape from the camp unseen. Some of those who succeeded in doing so encountered a stop group, positioned to the north of the camp, which picked them off. Meanwhile the 60mm mortar team was bombarding Chioco from where FRELIMO (Front for the Liberation of Mozambique) troops and ZANLA terrorists were firing at the SAS.

Sergeant Andy Chait's assault group approached the camp from the south. Crossing a gully via a makeshift bridge, he and his men moved through a field of maize until they came under fire from terrorists in a trench to their front. These were engaged with AK-47s, fragmentation grenades, and an accurately thrown white phosphorous grenade, which exploded in the trench. Those terrorists, not incapacitated by the burning phosphorous, were dispatched as they fled. While clearing the trench, Sergeant Chait and his men came under fire from an RPD light machine gun and shortly afterwards he was seriously wounded in the thigh, suffering a ruptured femoral artery. Enemy fire, including shelling by some 75mm recoilless rifles sited in bunkers nearby, prevented the SAS medics from carrying out emergency treatment until they had moved him to cover behind some buildings. A medevac helicopter was called and this arrived a few minutes later. Unfortunately, the medic's efforts were in vain, because Sergeant Chait died during the flight to Salisbury.

Shortly afterwards, the SAS withdrew, leaving a scene of destruction with at least thirty-eight ZANLA terrorists dead and a large number wounded. They made their way over a distance of a few kilometers to an LZ from which they were extracted by helicopter under cover of four RRAF Hunters. The operation had been entirely successful and the enemy abandoned the camp.

Border agencies—checking the fences between Arizona and Mexico.

LAW ENFORCEMENT AND BORDER AGENCIES

Borders exist to outline the territory held by each nation. In a few cases these borders are disputed between neighbors, but generally most are settled. As each country has its own laws and policy, some have become richer than their neighbors and so for many years wars have been fought as one country tried to possess the wealth of their neighbors by force.

In addition, some countries have relatively low taxation on certain items and this, in turn, causes some individuals to use this in order to make money—smugglers are a good example. In the main, smugglers have been replaced by drug cartels that wish to traffic their goods over borders in order to make exceedingly large amounts of money. Finally, we have international and state terrorism; these groups use borders in order to infiltrate a country or to seek a safe haven on the other side.

In order to prevent these incursions, most countries have border agencies whose task it is to stop any illegal infiltration by foreigners. These agencies have a whole host of methods, including regular patrols, ground sensors, and drones. In certain countries, trackers are employed to follow up and border breach and pursue the offenders.

The use of tracking within law enforcement agencies is directed mainly at border security and to stop the influx of illegal immigrants of terrorist using isolated areas to cross from one country to another. The problem of border security is an increasing problem as people from the poorer nations try desperately to move into countries which offer a much higher standard of living. Additionally, in many parts of the world, certain border areas are used by terrorist or revolutionary groups to pass unnoticed and carry out attacks.

While borders are patrolled in a wide variety of methods, many are mainly dependant on the terrain and the inhospitable environment. For example, large expanses of desert with no clear demarcation lines, no habitation and little food and water to sustain life go unchecked. The same goes for very large mountain ranges which are covered with snow the entire year round, or the territories in the Arctic Circle. In such isolation, where humans find it hard to survive, there is little point in expending manpower to check the borders. Where human habitation is plentiful the borders can see a lot of activity.

Author's Note: In 2005, Muslim militants in the South of Thailand detonated some sixty-five bombs in a period of less than six months. Some were aimed at schools, other railways stations and local police stations. It is suspected that many of the bombs were actually made in Malaysia and infiltrated over the northern border with Thailand. The attackers tended to detonate a small bomb to draw officials to the scene and then use a mobile phone to detonate a second bomb intended to kill or injure the security forces attending the scene. In many cases the attackers simply slipped back over the border into Malaysia. For this reason, security has been stepped up on both sides, with Special Forces being drafted into the area to carry out follow up missions when evidence of illegal border crossing has been found.

MOUNTAIN RESCUE

Search and Rescue (SAR) is an obvious place one would expect to find human tracking skills, yet the amount of training in this art is fairly limited, with search parties relying on local knowledge of the ground or the use of tracker dogs. The aim of any SAR is to locate, stabilize, and extract an people who have become lost or in distress. This means those lost in the mountains, covered by an avalanche, or lost at sea.

Most SAR teams are comprised of local volunteers who know and understand their own backyard; this makes it easier to understand where the casualty might be located. In addition, many SAR teams will have access to helicopter use either for providing aerial observation and final extraction once the casualty has been located.

Mountain rescue involve both human visual tracking as well as the use of search dogs.

HUNTERS AND GAMEKEEPERS

Hunters and gamekeepers have always used tracking skills, albeit in the role of tracking animals or birds. This skill continues and, while it is not related to human tracking, the basic principles are the same. Many game keepers or *'gillies'* practice tracking as a part of daily life and come to recognize *'sign'* as if it were second nature.

Hunters remain one of the last bastions of visual tracking.

The *'gillies'* use of tracking is almost always to define where an animal has gone in order that it can be hunted; today this is mainly used for sport. Yet tracking a heard of deer is the easy part, sneaking up on them can be sometimes impossible. Animals are far more aware of their surroundings than any human. They can smell and sense humans from a great distance and, trust me; they are always on the watch.

Hunting is a sport carried out around the world, normally in areas festooned with one form of wildlife or another. While hunting originated from the need for food, sadly today it is more for obtaining trophies.

ECOLOGISTS AND EDUCATIONISTS

There are many outdoor activity groups that remain keen on learning the value of human and animal tracking. Certain people have a yearning to learn the oddest of skills, especially those which involve the Great Outdoors. You find groups which like bird or animal watching, while others will take a sea voyage to observe the whales. This observation is the basis of all visual tracking as each—in their own way—provide an insight to both human and animal behavior.

Once again, modern techniques are employed when it comes to tracking animals. Everything from whales, foxes, turtles, penguins, and a host of other species are constantly monitored from some cozy office once they have been 'tagged.' In many cases, such as tigers, this type of tracking is done to help preserve the species and ascertain their actual numbers.

Author's Note: From 1998 to 2003, I personally taught human tracking to a wide variety of groups. In particular, there was an outdoor group from Belgium that came annually to learn and practice their tracking skills. This group consisted of mainly men, whose occupation ranged from office workers to policeman. They were extremely dedicated and worked hard for the two weeks of the course. Many became extremely skilled and were able to follow a track that I had made by simply walking through the bush and open ground without deliberately leaving any clear identification markers. Given that the terrain and weather in high summer is extremely difficult to work in Spain, they succeeded in most cases. The course notes can be found at the end of this book.

TRACKING TERRAIN

Any form of tracking will differ with the type of terrain over which the visual tracker is operating and in what context. That is to say, a hunter tracking a deer in the forests of Canada will have different priorities to that of a solider tracking an enemy through the jungles of Asia.

Conditions such as weather, age, climate, etc., will all have an impact on the sign which is needed to follow a quarry. As will the type of terrain; tracking in the snow is a lot different to tracking in the rain forest.

Desert terrains present wide, clear terrain; however, any spoor can be quickly covered by wind.

Desert Tracking

Tracking in the desert has several good points and several bad points. In the desert, for instance, your field of vision is vast and it is fairly easy to see where someone has trodden. However, the loose sandy conditions do not leave a clear imprint except in certain circumstances where the sand has higher clay content. While any quarry within a distance of several miles can be spotted in open desert, the heat haze can be a horrific strain on the human eye.

Direction of travel is fairly easy to establish if you have good maps, as no-one wanders aimlessly in the desert; they are always heading for some form of civilization. Temperature is another factor; by day the desert can be extremely hot and at night extremely cold. This tends to annul the use of dogs during the day, but they can be very effective at night. It is normally much easier to move at night as the stars supply a constant navigation guide. Additionally, in a military environment your security is far better in the dark. Be aware that sound travels over great distances in the desert.

The upside of desert tracking is the distance at which spoor can be seen; the picture above clearly illustrates this. The problem is this action works both ways and it is easy for the quarry to see the tracker in pursuit.

Another downside of desert tracking is the transient nature of any track left behind due to the wind. Once again, this will depend on the type of desert you are crossing over; some are permanently shifting due to the prevailing winds while other hardly move at all.

When tracking over desert terrain remember that the quarry will be heading for a specific location, as the survival needs of both man and animal are vital to life. Water holes and safe havens are few and far between, and their locations are well known. With this knowledge the visual tracker can forge ahead with a clear indication of where the quarry is heading.

Tracking across the desert is not as easy as one might think. While there are very clear marks in sand, it is almost impossible to read the footprint sign. Additionally, the sands move all the time. Where the desert turns to rock, the chance of seeing footprints is infrequent. The sun can be a good sign if it is in the right direction as reflection from any metal object can be seen for many miles.

In the desert it is essential to keep sand and stones out of your boots or shoes. It can be very frustrating and demoralizing to be forced to stop repeatedly to empty them. Stay covered up to avoid sunburn. While it is best to rest in the heat of the day and move at night across the desert this is not always practicable for a tracking team. However, it is possible to track at night using the moon and stars which supply not just light but a constant navigation guide.

Arctic Tracking

Many of the techniques that apply to the desert also apply in Arctic conditions—but in reverse. You need to dress according to the climatic conditions and pay particular attention to your feet and warmth. Survival is equally as important as tracking in such a harsh environment.

Moving over a snow covered terrain leaves lots of sign which are easy to spot from a great distance, added to which they are easy to track at night. This makesss no difference if you are tracking over oven tundra or lower down in timber country. Your quarry is going to be either on foot, which will make them extremely slow in deep snow, or using snow shoes or skis.

The main drawback to arctic tracking is the extreme temperature, which while helping to preserve sign, also makes it difficult to operate. Fresh snow falls will also cover any tracks quite quickly. As with desert tracking, the quarry will normally have a final destination in mind as it is hardly possible to sustain life outdoors for any length of time.

Shelter and warmth are critical to survival and therefore the quarry will seek the best avenue which will provide this. The visual tracker must anticipate this prerequisite and use it to gain ground and close the distance between himself and his quarry. One other important factor is the weather, as shelter is a real must as the day grows old and the nights become unbearably cold.

Jungle terrain is possibly man's best type of terrain to track over.

JUNGLE TRACKING

The one environment which lends itself to visual tracking techniques is the jungle. The reason is simple, as any quarry moving through the jungle must either stick to the local's tracks (most of which are single file) or head through the jungle forest. In either case, there will be lots of signs to follow.

It is hardly ever possible to travel in a direct line through the jungle. Most quarry will make use of streams or rivers, game trails, dry water courses, or ridge-crests. There may be native paths, which offer an easy path, but also make it easy for them to be followed. The alternative is either cutting away through the vegetation, which is exhausting and slow, or learning to move slowly without cutting, bending, and twisting your body so that it does not become entangled with the many spiky plants. Travel in dense jungle can be very slow, 1km per hour would be a very favorable speed and 5km in a day would be very good progress. Jungle paths and trails become the animal's highway at night—so do not use them outside the hours of daylight—stay in your camp area. Few humans move through the jungle at night, as the animal and natural hazards are too great.

There is an additional discomfort to the visual tracker moving through the jungle; from leeches, insects, and a host of nasty things like snakes and poisonous plants. These not only cause discomfort but they can distract the visual tracker from his task. While tracking in the jungle is fairly easy, in the military environment it is also prime territory for ambush and booby traps. Alien sounds and smells are also much more prominent in the jungle.

Urban Tracking

Most of the world's population lives in cities, towns, or villages. The majority of which are constructed of concrete or hard surface material which leaves little or no footprint behind. In this concrete jungle, both human and animal tracking prevails; not only does it prevail; it is carried out on a massive scale. Where we once used the human eye and instincts to track our quarry, we now use highly sophisticated surveillance technology.

Urban Tracking— there is far more tracking done in an urban environment than any other type of terrain.

Literally billions of CCTV cameras around the world monitor a wide variety of tasks, from traffic to street crime. Computers cross-reference data while government agencies communicate and exchange intelligence on a terrorist group. Those that are special interest get the full treatment with their cars, homes, and offices being bugged while the individual is constantly tracked by one device or another.

If you look deep enough we track just about everything from credit cards, parcels to mobile phones—there is no escape.

SUMMARY

Visual tracking in its true form is a primeval skill which has changed little today. As with so many of man's earlier skills, they have been taken over by civilization or discarded for more modern electronic techniques. Yet we all continue to use some form of visual tracking in our everyday life; as we walk to the shops we avoid other people in the street, we have a path to walk and a destination to get to—in essence we are visual tracking.

No doubt there are still some isolated tribes in South America, New Guinea, or India that survive due to their visual tracking skills. Even the Iban of Borneo still practice hunting in the jungle forest, despite having televisions, fridges, and sending their children to school down river every week in long boats.

Then there are the dedicated few in America and Europe—mainly retired soldiers—that maintain the skill of visual tracking for their own enjoyment or hunting.

When it comes to successful visual tracking it is not enough just to follow the sign left behind by the quarry. These signs must be interpreted in order to gain an understanding of the quarry's movements in order to foresee their final destination.

One thing is for certain: there are few people on this planet that can hide away and never be found. In our everyday life we leave a trail so clear that anyone with the right access or equipment can easily know your rough location to within a few meters.

Chapter 2

UNDERSTANDING TRACKING

It is difficult to explain in a book how to learn visual tracking, as it is a skill that requires a lot of outdoor work with a simple explanation of what you are looking at. I remember someone giving me a piece of advice when I started writing and I think this applies equally to visual tracking. That is: Write as if you were talking to a blind man. Explain to him everything you see, hear, smell, and the environmental surrounding of the subject matter. Do not just say there is a woman sitting on a park bench. What time of year is it? What is the weather doing? What clothes is she wearing? Is she just sitting there? What look is on her face? Anticipation? Happiness? Distress?

In order to understand visual tracking from a book, you need to adapt the same outlook as talking to the blind man; fully understand what it is you are seeing, hearing, or smelling, as well as what you are feeling. So place yourself in an outdoor scene and survey the countryside, then look down and see what is a few feet away from you. If you are on a tarmac road, then you will not see many tracks; but if you are on a dirt track, you might see tire tracks or even a footprint. All you have to do now is interpret what you see into usable intelligence. Tire tracks mean a vehicle has gone this way, while the footprint will indicate a human—simple, but you have to push this a little. The best example I can offer is one of returning home only to find the door you had previously locked, slightly open. You would immediately look at the door for signs of forced entry. If there is none, who else has a key to your home may enter your mind? Pushing the door open, you see everything as it should be, there is music playing, even so you are cautious. Why? Because you have seen signs and it could be a burglary still in progress. Relief as you discover

Visual tracking sign—footprint are important because this is where man makes contact with the earth.

your significant other home early from work—in fact, you had just accomplished some visual tracking skills; you spotted the signs and evaluated them to what could be the cause.

This is the approach you need to adopt with visual tracking: interpret what you see. Nevertheless, before we can start, we need to know what 'signs' there are out there, what triggers will help us understand what we are seeing and finally, how to interpret them.

VISUAL TRACKING SIGN

While tracking starts with a simple sign of our quarry which we are able to track due to the prominence of the sign, in reality it is a precise art which requires many years of skillful practice to achieve. In a military situation visual tracking is fraught with additional dangers as the enemy may wait in ambush, lay mines, or prepare booby traps. Despite this, recent history has shown that with the right skills, experience, and a large amount of fortitude, a quarry can be tracked providing the tracker has the following attributes:

- Good knowledge of the local terrain.
- Steady and patient.
- The ability to move slowly and quietly.
- Detect and interpret all signs.
- Avoid ambush and booby traps.
- Understand the weather, both before and during any tracking.
- Know when the track is lost.
- Keen sense of sight and smell.
- Know the quarry and its habits and behaviors.
- The ability to listen to their own inner voice and heed any warning.

The last item in the list is possibly the most important; man has the ability within him to 'feel' through the use of senses long since forgotten. Our ancestors had these senses which, when mixed with a basic knowledge of the terrain, animal habitat, etc., allowed them to become successful hunters. A good tracker must learn to rely heavily on these old senses and while not understanding them, at least listen for them. When tracking a quarry, you should build a picture of it in your mind: What am I following? How are they moving? Do they know I am following? This picture will develop as you progress along the track; colored in by the signs left behind. A simple paw or footprint in the dirt will provide so much information: first and foremost, it provides confirmation, identifies the quarry, indicates direction and provides an experienced tracker with a rough time when it was made.

To understand any form of tracking, you first need to understand all the essentials that make tracking an art. If you understand these, then you are halfway to becoming a good tracker. Study all the items in the list and become familiar with each, because as a whole, these are the items that make up the tracking picture which you will eventually follow.

HUMAN VISUAL FIELD

Human eye detecting small detail is what visual tracking is all about. The bayonet is from an AK47 (most common weapon of terrorism) stabbed in the ground with its scabbard placed nearby. This is the kind of detail you need to see.

As most tracking is done by using the eyes, it is good to have an understanding of how we see as humans, as this is vital to spotting and reading the signs. When light reflected off an object enters the eye through the pupil, it is flipped upside down by a clear lens behind the iris. The light is then projected onto the retina, the part of the eye where vision is generated. The retina contains a lot of vision cells, of two main types: cones and rods. Cone photoreceptors are most important to color vision. The cones distinguish color based on the light's wavelength. For example, as covered earlier in this book, a long wavelength is red and a short wavelength is violet. The photoreceptors then convert the wavelengths into electrical signals that are sent to the brain, via the nervous system, creating color sight as we know it.

When tracking, you need to understand what is happening with your eyesight and, where possible, work with a smooth head-turning motion. When you look straight ahead, the middle of your eyes do not register the same as the sides of your eyes, as the sides are more sensitive to light, brightness, and any flickering movement. (How often have you said: *"I saw it out the corner of my eye"*?) The next time you are in the cinema, try this: Look up to the ceiling while the movie is playing—especially during bright scenes—and you will clearly notice that the movie flickers because you are seeing how many FPS the film is running at.

If you were placed in a blacked-out room for two hours or more and suddenly there is a very bright torch switched on would you see it? Yes. However, what if the torch was only switched on for 1/200th of a second? The answer is still yes, as the contrast will be registered on the brain. Tests carried out by the US military similar to the above showed that a pilot could recognize and identify an aircraft type flashed on picture for just 1/220th of a second.

Another thing about your eyes is color; we see green best, which is why at dusk all the greenery and vegetation look so beautiful and bright with all the green color enhanced. This is because we see green clearer than say blue or red.

So how does this help the tracker? First look directly at the scene in front of you, using the center of the eye. Move your head slowly from one direction to another. Look through the vegetation—not at it. If you detect something and are not sure, move your eye to a closer spot for a second, then look back again and let your eye re-focus.

Author's Note: While not connected to tracking, it is worthwhile pointing out a similar tip. If you are shooting at any range, just before you pull the trigger, let your eye fall to look at the green grass close by for a second, then re-sight and fire as your target becomes clear. This is an aiming skill that I was taught years ago by a Russian soldier from their Sniper School, who was visiting UK and I have applied it to tracking.

LIST OF ABBREVIATIONS

Normally the list of abbreviations would be placed at the back of a book, but in the case of human tracking, it is essential that the reader understands some of the terminology relevant to tracking in order to comprehend the techniques used. For example, would the reader normally know what 'spoor' or 'top sign' meant without a simple explanation?

The brief description below will aid the reader in comprehending all the facets that lead up to the interpretation of a tracking picture. At this level of ability, the tracker will be able to identify his quarry with great accuracy and follow them at a speed where they can catch up. This is the same for both human and animal tracking. While some books describe the same aspects of visual tracking, other will vary slightly; the important thing is to recognize the sign for what it is.

Quarry: The person, animal, or vehicle being tracked.

Tracking: This is defined as the skill of being able to identify a specific person, animal, or vehicle by the signs they have left behind. It is also about making an intelligent interpretation of the sign to enable a prediction and history of the quarry.

Sign: Is any physical indication made by the quarry during their passage. This can include visual sightings, sound indications, or smell. The sap from a bruised root, disturbance of wildlife and birds, lack of dew on the early morning leaves. All this can be evidence of passage.

Sign is divided into various parts:

Ground sign: A clear footprint in the wet sand would be classed as ground sign.

Bottom sign: Is generally anything below knee height and would include short grass that had been pushed in one direction.

Top sign: Anything above knee height, a broken branch where someone has grabbed for support.

Direction of travel: The probable overall route taken by the quarry. Assessing the direction is a major factor in any tracking operation. This can be deduced not simply by the direction of the spoor but also with knowledge of the quarries habitat or end goal, such as an enemy heartland. Terrain type and accessibility over the terrain such as natural obstacles, i.e, large rivers also play a major part in the evaluation of the direction of travel.

Conclusive sign: The butt heel shape of a machine gun found in an old camp site would be a conclusive sign that the quarry was armed.

Key sign: Where there are several types of signs along the same track, the tracker must be able to identify something that is 'key' to the quarry they are following. This could be a particular tread pattern, or the stride measurement of the quarry. The last person in a multi-human track normally leaves the clearest footprint: use this as your 'key' sign.

Substantiating sign: Anything that can be seen, heard, smelt, or felt which would substantiate other signs in close proximity.

Temporary sign: Any sign that is removed or degraded by the weather and time. The tracker must be able to identify when the sign was made and what time passed to when it was located. Local weather conditions and the ground type will play a large part in this deduction.

Time Bracket: All sign is made at a certain time, then due to erosion—weather or other mitigating circumstances—the sign will wear down. Upon locating the sign for the first time the visual tracker should place a time bracket on the sign from when it was made to its presence state.

Straight edge: Most humans and vehicles are heavy and therefore, apply pressure on the ground with their feet or tires. Under certain conditions the imprint leaves behind a clear straight edge of the sign.

Cast: To look around in a logical pattern from the last known confirmed sign.

Diet: Relating to the excrement of either human or animal to ascertain their dietary habit.

Entry and Exit: A person being tracked has to make a choice of where they enter and where they exit, under certain conditions, the tracker can visually determine these points.

Environmental Medium: This happens when the quarry is being tracked through a forest which suddenly stops and turns into a ploughed field. Where the visual tracker had been looking for scuff marks on the forest floor them will now look for footprints in the soil.

Environmental: The sun, rain, snow, and wind all help to degrade or change any sign left behind by the quarry.

Habitat: The place where the quarry (both humans and animals) normally live. For example: the Iban tribesmen live in the jungle of Borneo.

Pace: Generally refers to the speed at which the quarry is moving, which can be determined by the length of the pace.

Permanent sign: A broken tree branch would be a good example.

Physique: The size and type of quarry being tracked.

Pointers: Something that indicates the quarry may have gone before the tracker.

Spoor: Sign left behind by the quarry.

Starting and Defining: A clear indication of the quarry and what it is. Soldiers who have initiated an ambush, fired ammunition and then ran into the bush, would be a good example.

Static Tracking: Waiting unnoticed, on a known route taken regularly by the quarry in order to establish presence and confirm spoor.

Step: The pattern of the quarry made by its movement, i.e, a footprint or animal hoof print.

Track: The route of the tracker following their quarry.

Tracking picture: A combination of track sign spoor which builds up a picture of what the quarry is, where they are heading, and how many there are in the party, etc.

Scent: This is normally refereed to when using dogs to assist in tracking, but it can also apply to humans as smell carries in the wind in some climates.

Deception: This is where the quarry is aware of a possible tracker following their spoor and makes every effort to leave behind little or no sign. They also try to falsify signs to mislead the tracker.

Track Jumping: This is where the tracker has a clear indication of the route and jumps ahead at full speed to gain on the quarry.

Character: Most human quarry leave behind some form of individual trait that identifies their character. For example, soldiers often wear a typical type of footwear. The quarry may have worn tread on their boot pattern or a stone may have become lodged in the tread.

Key sign is any individual makings that will indicate an individual quarry.

TUNING INTO THE QUARRY

From the first 'key sign' and any other sign the visual tracker might deduce from their findings will help them 'tune in' to the quarry and assist the visual tracker by helping them predict what actions the quarry may take.

HOW AND WHERE TO LOOK

Search the ground about three to four meters in front of you. Move your head slowly, side to side, while concentrating on what you see. Look right through any vegetation to the ground. Tracking sign is not always easy to see; even when it is obvious and sometimes you will see lots of sign and then suddenly nothing, for several meters.

As previously mentioned, humans will take the easy route and unless trying some form of deception, will stick to the overall direction of travel. Always check for signs at the most likely places where signs may be made:

- The edge of a clearing.
- Where you either enter or leave an area of vegetation for an open area.
- Near the banks of rivers and streams.
- Soft dusty areas.
- Muddy areas free of vegetation.
- Where thick bush has barred the way of any quarry.

Where to look— when tracking over desert terrain a good place to look is near any natural water catchment as they are few and far between, and everyone needs water.

What to Look For?

Now that we know some of the main elements of tracking, we also need to understand what it is we are looking for. Who or what are you following? How long ago did the quarry pass this way? What is the weather like? Do we know where the quarry is heading? These are just a few of the basic questions that must be answered before we can start. When we have the answer to these as a starting place, there are two main physiognomies an experienced tracker will look for, transformation and discoloration:

Transformation

Discoloration, transference, key sign, and much more can be deduced from a boot that has recently stepped in water and transferred it to another medium.

Transformation can be anything that has changed from its original state. Where there was soft soil, there is now a footprint; where there was a branch, there is now a broken branch. In some cases the transformation offers little information, like a broken branch for example. By comparison, a footprint might be made by a boot, barefoot, or a paw, from which much detail can be assumed. Transformation can also be a continuous thing, where one footprint has trodden on an older footprint, thus aging the original.

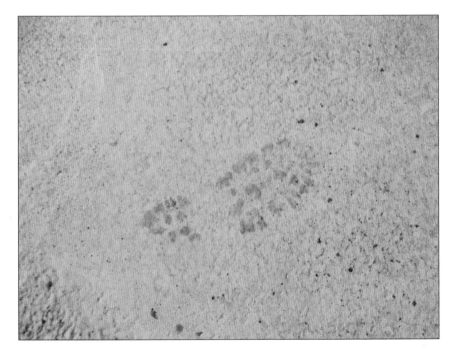

Discoloration

A turned stone is both displacement and discoloration; in some terrain this is easy to see, while in others it is not so prominent. This applies to a broken branch which has been damaged for several days, as the leaves will have changed color. After crossing a stream or stepping in water, there will be discoloration. Additionally, the quarry will leave behind some discoloration by depositing wet prints on the ground. Depending on the terrain and prevailing weather conditions at the time will determine how long this sign remains visible to the tracker.

Some discoloration is easy to detect; yellow snow is an indication that the quarry has urinated. Where the tracker is following a wounded animal or the enemy after a contact, there may well be blood. This is of great help to the tracker as the discoloration of blood against the natural terrain is quite clear. It also provides quite a lot of information. A wound from the lung will be bright and frothy, whereas blood from the lower body and groin will have the appearance of a dark, tar like substance. The volume and frequency of blood is also a good indication as to how wounded the quarry is and, therefore, to specify the speed at which they are moving. To the

Footprints, paw prints, tire tracks, all going in different directions. Some have stepped over older prints, but all help provide a tracking picture.

experienced tracker and with the wind in the right direction, fresh blood also smells, so alerting to the tracker that the quarry is close.

Other factors such as terrain, weathering, litter, and possibly camouflage (if it is a military track), will also make a great difference to the quarry track. The tracker must be proficient in understanding the effects of rain, wind, sun, snow, and time on any sign left behind by the quarry. If there has been a military engagement, then the tracking force must be aware of the possibility that the enemy may try to camouflage their tracks. In most cases, the fresher the track, the easier it is to follow as time allows the spoor to become eroded by weather or diluted by others covering the quarry's spoor.

Footprints

Please note that, to stop any repetition the word *'footprint'* refers to both human and animal paw prints.

As people normally relate a footprint with tracking, we should look in detail at what this sign offers in the way of information. First, there needs to be a starting point, which is normally a place where there is sufficient evidence of our quarry, that is to say a series of footprints. This we normally refer to as the start point. The best place to ascertain a start point is one where the quarry has walked over a stretch of soft clear damp soil or any terrain which acts as a medium, which will produce several clear *'casts'* of the footprint. The more clear casts we have, the more information we can interpret.

So much can be gleaned from several footprints, such as how fast the person is walking, if they are wounded, male or female, and whether a person is carrying a heavy load or not. We tend to look at a footprint but not see anything other than the type of sole of the footwear the quarry was wearing. Yet there is so much more that we do not observe:

- The length of the footprint from heel to toe.
- How wide the sole is across.
- Distinctive marks on the imprint, maker's brand name or a chip out of the material, or a stone wedged in the tread.

The Style of the Footprint

Where this is more than one footprint in succession from the same quarry, then we can measure pace and stride.

Let us look at some of these points in detail. For example, try walking yourself on a dirt track: take five to ten steps then turn around and walk backwards in parallel to the first footsteps. Look at the difference and note the changes.

Next find a clear spot and walk five to ten paces . . . this time dragging your right leg as if wounded. You will find the left footprint is deep as your muscles compensate for not putting pressure on your right leg. You will also see that the right leg is slightly out to one side as the body naturally swings the leg forward again in an effort to relieve any pressure. The same pattern but with another set of footprints very close, with a short step could indicate that the injured person is being assisted by another person.

Someone walking with a limp or a sprained ankle will favor the injured foot by placing only the tip of the foot down in a kind of hopping motion.

Measuring the stride of your quarry from heel to forward toe will provide a pace. It is best to get yourself a stick and

A comparison of footwear tells us so much—the larger one is a heavy male around 210–225 pounds, while the smaller one is a young lightweight person.

break it so that it fits the pace of your quarry, this way you do not have to keep stopping to measure the stride.

If the stride widens or shortens then it can reveal several clues. A shortened stride might mean the quarry is getting tired, while a wider stride may mean they have upped the speed they are travelling. The stride of the average middle-aged male is just over a meter, while the same man running may well increase their stride to 1.5 meters or more. The faster we run, the straighter our footprints become and at an all-out run, there would only be the sole showing as the heel will not hit the ground.

By comparison, anyone carrying a large burden will have a short stride and the footprints will mostly point outwards as the body tries to maintain balance.

One of the best ways to learn the difference is to practice on a flat dusty road or track. Get a branch and clear any previously made footprints or vehicle marks so that the surface will display only the footprints you make. Walk deliberately without stopping and then go back and measure your stride, break a stick to that size. Now run as fast as you can parallel to the same set of prints, again use your stick to see the difference in stride and the impression made by your foot.

If you have a partner of the opposite sex, then try doing the same footprint test together walking side by side. You will find that it is fairly easy with a little practice to distinguish the difference in sex. Do not just look for the style of

Direction of travel is one of the first things a good tracker must establish. Both man and animals take the easiest route—in this picture it is fairly safe to surmise that they went down the valley bed, and did not climb the steep sides.

footprint, look how deeply each has penetrated the same soil; most (but not all) women have a slight turned-in toe within their step.

Once you have done that, try a fireman's lift with a friend and then walk over the ground again and observe the footprints. Finally, look in detail at all the footprints you have made and make a note of the difference. It is a good idea to use the camera on your phone to take a few images you can use for reference later.

DIRECTION OF TRAVEL

Direction is possibly one of the most important factors of visual tracking, as it is an indication of the overall route the quarry is taking. In a military situation there will be intelligence that indicated which way the enemy would normally be heading, but if we were tracking an animal that is grassing it could wander anywhere. With a good knowledge of the terrain and what he is tracking, a good tracker will be able to determine where the quarry is headed and push forward quickly. Judging the direction of travel of a quarry is vital to speed and interception, especially in a military operation.

Assessing direction is a major skill when visual tracking, especially when the spoor is over hard ground. There are normally lots of indicators as the quarry's direction of travel and if they diversify, then there is normally a very strong reason, i.e, a river in flood or another obstacle that is dangerous to cross, thus forcing the quarry to leave the direction of travel. In most cases when this happens, the quarry will regain the direction as soon as possible.

A simple direction could be established if you are tracking a quarry through a valley with very steep sides. The odds that they will climb a steep cliff or scale a very high mountain are limited if the quarry has no indication you are following.

When thinking about direction, the tracker must put themselves in the mind of the quarry. Where are they going? In a guerrilla war it is fairly easy to establish as they will most probably head for safe territory. A lost hiker (if they obey the rules) will head downhill and someone going from A to B will hopefully have told someone where they were heading.

Start Point

When doing your first few tracks use a simple 'Start Point' such as a track through a forest where you are certain to locate a clear set of footprints.

Start points are extremely important and a good tracker will locate one quickly. In a military situation this is fairly easy if the enemy has made contact and had a fire-fight or has laid in ambush. There will be lots of signs for this, such as empty ammunition cases, etc. The enemy will also run for safe location after the fire-fight or ambush and the tracker will know this. It is, therefore, a simple matter of following the enemy until the tracker finds a soft spot of ground which

has lots of footprints. Another good start point is where the enemy has spent some time or rested overnight.

A good start point will indicate the number of people in the party (this is a simple matter of counting the different footwear or sizes over a given area). It will reveal the sex of the individuals, as women normally have smaller feet and a shorter step than men; if they are carrying anything, because this deepens the foot imprints and may indicate load carrying. The length of a stride will indicate the speed at which the quarry is moving.

Moreover, a professional tracker will also look for individual footprint identification markings. It is not unusual to have several different types of footwear styles in a party. In this case it is best to try and identify the strongest print, as this is normally the last person in the group.

Starting Point Indicators

It is generally accepted that the visual tracker with the most experience over the type of terrain they wish to track will make an assessment of where to look for the best starting point. In many cases this is done for them as they may well be tasked to do a follow-up after an IED incident, or where an ambush has taken place. In this case it is a matter of establishing the direction of travel the enemy will have taken. Where there is no firm evidence to indicate a start point, the tracker will use his local knowledge and check that area for sign. These include:

- Traditional areas where the enemy have appeared.
- A known or most likely campsite.
- River banks, and known crossings.
- Areas of track major junctions; expressly MSRs.
- Areas of steep gradients—especially in the jungle—as a climbing man leaves a lot of signs.

In many cases it is the skill and local knowledge that will allow the tracking leader to locate the direction an enemy has gone.

Where no evidence is found it remains to carry out a search of an area, similar to the *'cast'* when the track has been lost. However, the cast area will be much larger and depending on the terrain may well turn into a type of *'fan'*

Locating an enemy camp will tell you how long the quarry stopped there. How many there are, if they cooked food, the state of their moral, and a whole lot more.

search using all available trained trackers to spread out and look for a start point. While this method is time consuming and will require covering a large area of ground, it can prove effective. A normal fan search will start from a chosen spot and cover distances up to 200–300 meters. If there is a track in the vicinity, this may well be used to form a baseline; or a similar action can be carried out using a river or stream. This type of search becomes more difficult when carrying out a cross-grain search.

Enemy Camp

Finding an overnight camp will reveal a lot of information to the tracker—especially in a military situation. It will also act as a good starting point. Litter and poor camp discipline is one of the great benefactors to the tracker. It offers much information about the quarry and, above all, it provides confirmation. A quarry which has poor discipline and leaves litter along the trail is making life very simple for the tracker, as it allows him to jump ahead very quickly. It also tells the tracker that the quarry is not caring if it is followed or not and therefore, they will be noisy. Cigarette smoke and the sweetness of chewing gum will be detectable for up to 100 meters, more in a close jungle environment. The silver wrapper from a chocolate bar can be seen up to 300 meters on the floor of the desert, while human and animal feces attract flies, thus drawing the tracker's attention toward it.

Poor camp discipline will show signs of fire, sleeping, cooking, and possibly weapons. The experienced tracker will derive so much intelligence from one poorly disciplined campsite used by the quarry (see Chapter Five).

Seeing the natural entry and exit points will allow you to push forward quickly—no quarry in their right mind would push through thick gorse and thorn when there is an easier route.

Natural Entry and Exit Points

Where there is an indication of deliberate sign, the tracker will look for natural entry and exit locations and, in doing so, will find a good start point. This theory is best described by our normal attitude of everyday life.

When we enter a room we generally go through the door, even though there is a window we could climb through and a wall in which we could break a hole—we do so in the understanding that the door is the logical choice of entry and exit. If the door becomes jammed and we were trapped in the room, more than likely the second most logical choice of exit would be the window. If we were a prisoner of war or in jail, we might consider trying to dig our way out or breaking the wall to aid our escape . . . but this would only be in the worst case scenario.

The same natural choices are presented in tracking. If we enter a forest which is comprised of many trees, we do not climb up and down the tree, we go around them taking the easiest route that will lead to our destination. The same goes in the jungle or a city for that matter: we take the logical, easiest route, which will get us to our destination the quickest. Therefore, a good tracker will simply look at the most logical direction of travel of the quarry and select a route based on the natural entry or exit points.

THE EFFECTS OF WEATHER

Sun, rain, wind, snow, and freezing weather conditions can all affect sign in as much as they will change the appearance of the sign or make them disappear altogether. The time span over which the change takes place will be dependent on the severity or type of weather conditions. For example, a fresh fall of snow might only half conceal previously made footprints; likewise, the sun may dry temporary sign left by a wet print after leaving water. The major effects of the elements on sign are:

The effects of sun on old tracks as seen here where the sun has baked the tire tracks and time has cover them with pine needles.

Sun

The sun has a drying effect. In some cases this may be helpful or detrimental to the tracker, depending on the effect. The sun will help preserve a wet footprint in clay conditions, while drying out sign that could be left on rocks or other solid surfaces. A broken branch will also be more prominent as it will have dried and changed color, making it more notable than what it was when initially broken. At the same time, some grasses that have been trodden down may well recover to some degree.

Rain

Rain will melt snow, wash away a clear footprint, and generally deface most types of sign. Rain is not good when you are tracking. However, if you have the direction of travel clearly defined, the rain will help you move ahead as it will mask any noise you may make.

Snow

Tracking and identification in virgin snow is fairly easy. If you also know the time of the last snow fall you should be able to estimate how far ahead your quarry is. Fresh snow fall will help cover the spoor, but it takes a lot to fill deep footprints. The degree of snow fall will alter, depending on if the footprints are out in the open or in a forested area.

Wind

Depending on the terrain will determine how much effect the wind has on sign. The wind can move leaves, drift snow and sand. It can also help reinstate trodden grass and dry out wet footprints. Over time, wind has an eroding effect on all sign.

Freezing Conditions

A footprint made in wet ground can remain for any length of time if it is frozen. Freezing preserves sign to a certain degree, especially in wintery conditions.

DECEPTION

There are many other factors that affect tracking, one of which is the awareness of the quarry. An animal grazing will wander aimlessly looking for the best food, while a human enemy soldier might wish to break contact and make to safe ground as quickly as possible. If the quarry becomes aware of your presence, they will take action; a wild animal getting scent of your approach will almost certainly run, while a human may try to deceive you.

Many people have tried to deceive a tracker by frequent turning around and walking, thus making backward footprints. This could also mean that someone in the party is constantly looking backwards to see if they are being followed. In this case the tracker will spot the deep, short heel prints which are a clear indication of someone walking backwards in an effort to throw the tracker off the spoor.

A professional tracker will spot these, as someone walking forward does not simply start walking the other way. So he will ignore these and continue to follow the reverse footprint for some time, in the belief that the quarry will once more turn in the original direction.

Using a branch to rub out or obliterate the sign is also fairly easy for a professional tracker to spot. The tracker will suddenly see no sign, but if he looks carefully, he may well see half a footprint with the other half brushed over. If the deception has been done in haste, there may also be evidence of the bush marks themselves. In such a case, the tracker will look for the obvious exits and check that the quarry has not chosen an exit off the route.

The tracker will immediately know if the spoor has gone cold. Where there is no sign at all, a 'cast' will take place from the last known confirmation sign in order to re-establish the spoor. This sweep will generally take the form of walking in a circle some meters from the last known sign. If nothing is found, then the tracker must rely on his judgment and follow the original direction in the hope of regaining the track.

Deception techniques are a clear indication of the enemy's intentions, which will provide the tracker with useful information. It will also make the tracker very aware that the deception may be followed by the enemy placing a mine, IED, or an ambush. It is also a clear indication that the tracker is on the right spoor, but needs to be careful.

SUMMARY

You should now be aware of visual tracking signs and the categories they fall into, added to which you should also understand a little about what they mean and how they are affected by weather, time, and environmental conditions.

You should also be aware of where to start looking for sign so that you have the best start point possible and the clearest footprint.

Finding this start point is vital to any visual tracking as indicated in the true story below.

During the Vietnam War, Sergeant Lawrie Fraser was in command of a patrol from C Troop 1 Squadron SASR during its tour of duty from March 1967 to February 1968. Although Fraser was only twenty-nine years old at the time, he was well experienced in jungle fighting and an excellent visual tracker. Having joined the army in 1956, he had served in Malaya and later having joined the SAS, went on to patrol in Sarawak.

On the evening of May 3, 1967, a helicopter dropped his patrol into an LZ five kilometers north east of Nui Thi Vai, in the western part of the Phouc Tuy Province. The helicopter had dropped the patrol in a small open area, covered with meter-high grass. This grassy area was no larger than 100 meters square and surrounded by high jungle, but almost immediately, Fraser found some fresh tracks. The patrol started to track the enemy, but within minutes, the patrol came under fire. It was answered with suppressive fire from the air as the helicopter fire team was still covering their insertion. Fraser realized that their position was now hopeless due to the compromise and immediately recalled the helicopter to extract them.

On the afternoon of May 18, the patrol was in action once again; this time their mission was to carry out an ambush north west of Binh Ba. Once more, they were inserted into an LZ by helicopter where everything seemed quiet, so they set off for their objective. Not far along the route, they noticed 'sign' and started to move forward slowly. Some half hour into the follow they spotted a single Vietcong (VC) soldier. Private Jim Harvey immediately fired off a 40mm grenade and the VC was killed. Suddenly a heavy machine gun, manned by a group of VC, opened up on the patrol and Harvey was wounded. Taking Harvey with them, the rest of the patrol tried to get back to the LZ under the cover of smoke, putting through an urgent request on the radio for an extraction. With the VC attempting a flanking movement to cut them off, it was a race as to who would get there first. Fraser's patrol won. Once again, a helicopter light fire team laid down suppressive fire, although this time the fire fight was much fiercer than at Nui Thi Vai. Eventually, under heavy enemy fire, but without any further casualties, the patrol was extracted.

Emergency extractions were not always necessary, however. On June 7, Fraser's patrol was again dropped off at an LZ in an area of primary jungle seven kilometers north east of Nui Dat. This time the insertion was completed without any problem and they set to tracking down the enemy. Two days later they came across clear, conclusive sign of the enemy and due to the numbers, decided to lay an ambush of Claymore mines on a track frequently used by the VC. Sure enough, at 14:45 hours, six enemy troops walked into the ambush and were all killed. One happened to be a VC tax collector and the patrol found a large sum of money on him. Their mission a success, the patrol was extracted and returned safely to Nui Dat with the tax revenue.

Chapter 3

BASIC TRACKING SKILLS

There is only one way to learn about visual tracking . . . and that is to actually do it. Reading books will only tell you what to do; in essence you really need to see and feel visual tracking. The previous chapter highlighted most of the various sign that visual tracking is built on. It is now time to do a little outdoor work and do it for real. To fully understand visual tracking you need to start off by using a place where tracks and tracking sign is plentiful. Most people can find themselves a small path through a park or forest, but if you live in the city it is best to drive out to the countryside. You are looking for a dirt track, maybe one that is walked by people on a daily basis, but best to choose one that is not too popular as too much footfall will only confuse you. It is also best that you do this with a friend or in a small group, as one can act as the quarry while the others follow. Before you start, make sure you obey a few basic safety rules:

When tracking, observation is all about seeing what was there by the sign left behind. In this case an AK47 assault rifle was rested against a rock.

- Always let someone know where you are going and roughly how long you will be.
- If you have a mobile phone, take it with you (one with a camera is best).
- Check the local weather and dress for the conditions.
- A map and compass, plus a notebook would also come in handy.
- Use good footwear, and check to see if it has a pattern on the sole.

Now that you understand all these facets that govern tracking, we are going to start learning how to actually find these features and interpret them for what they are. I make no apologies for starting simple and building the picture up slowly. We will start with observation as this is the foundation of all tracking—understanding what it is we are looking at and interpreting its meaning.

OBSERVATION

All tracking involves observation: seeing and understanding what you see. As explained in Chapter Two, the human eye is a very complicated organ and you need to understand how best to use it. There is nothing better than setting up and doing a few observation tests.

My personal favorite is to get my students out in the countryside and walk them down a lane where I have placed a number of objects. The first time we just walk and I say nothing, but at the end of the trail I ask them what objects they observed. Some will have made a mental note of seeing some of the objects, others will see nothing. Next, walk them through, but this time make sure they all make a mental note of what they see with regards to an object. All the objects should be manmade, such as a cigarette packet, tin drinking mug, matches, candy wrappers, and so on.

Author's Note: Most items could have been dropped on the ground naturally by others and while alien to the surrounding vegetation, they may or may not have meaning to the tracker. For this reason I try to stimulate the student's mind by dropping items that are totally out of place with the natural surroundings, for example:

- Fish bone skeleton
- Ladies' lipstick
- Watch
- Pen
- Coin
- Handkerchief
- Pocket knife

Next, we simply place a series of similar objects away from a vantage point at various ranges starting from a few meters out say to 20 meters. The human eye is extremely good, but not good enough to detect small objects such as a matchstick at little more than 20 meters. So place your items accordingly with the small ones closer to the vantage point and the larger ones further way. Make some of the items colorful, some with shine or color contrast, while some should be in direct conflict pattern wise to the natural surroundings, etc. Now ask your students to make a note of what they are seeing.

If the students do not see the objects and items you have placed after a given period of time (ten minutes is usually enough), then you should point them out. It is also a good idea to run a quick observation exercise most mornings using the same objects so the student understands what they have seen and what they need to look for.

The upturned rock is easier to see than one would suppose, it just takes a little practice.

OBSERVATION EXAMPLE: UPTURNED ROCK

Many people think it's not possible to see the difference if one rock is upturned among a floor covered with rocks; they would be wrong. One upturned rock is fairly easy to spot, as the dryness of the earth beneath the rock will also help indicate the time the quarry passed through the area. Any sun will dry the soil and the base of the upturned rock;the darker and wetter the rock, the closer your quarry. Look closely at the picture and see how easy it is to see the upturned rock center frame.

FIRST TRACK

As previously explained, the best way to start tracking is to find a local area used by walkers and hikers close to you. This may be the local park if you live in the city, or a nearby forest if you are lucky enough to live in the country-side. The general idea is to start off simply and expand your knowledge and understanding of the basic tracking skills.

You have to encourage people to become visual track-ers. If the first attempt at tracking is too difficult, then you will demoralize the students and they will give up. Always start off simply and make your track very easy to see sign and therefore easy to follow. If you are going to leave footprints, make sure you do so in soft soil so they are clearly visible.

If you are instructing someone, then explain what they should be seeing as you walk with them. Always stay a little to the rear and let the students be the front-run-ners. If they miss something, stop, and show them so they will remember to look harder next time.

A simple forest track is an ideal start for your first track exercise.

EXAMPLE OF A FIRST TRACK EXERCISE

The image above is of a local forest close to my home in Spain which I walk by on a regular basis. This affords me the luxury of demonstrating the many facets of tracking and what information we can derive from following a quarry. I make no apologies for the simplicity of the exercise, as I am sure this will allow the reader to fully understand the points I am trying to teach.

The point I have chosen to start is the entrance to one of the walks taken by many of the local population. It is a pleasant walk through a tree lined forest. The path itself, while mainly covered in embedded small rocks and hard packed earth, has many soft soil areas where footprints can clearly be seen.

My aim is to allow the students to see the footprints, note any identification markings on the sole and be able to follow and identify a direction of travel. From all this, and some good map reading, I want them to identify a possible destination.

For those who have never undertaken any form of visual tracking I would ask that you carry out some explanation as to the basics. The best way is to build up a series of simple PowerPoint presentations. To help you in this you will find an example on 'Sign' at the end of this book which you are free to copy and distribute.

The Start Point

Here we have a good stretch of soft brown soil which is covered with many human and animal tracks. I am going to select just one footprint and, naturally, I have chosen the strongest, clearest sign.

When there is lots of different sign, pick a prominent one as your starting point.

Looking at the area, it is clear that many people have walked both ways along this track, but the one I have chosen looks to be relatively new.

> **Author's Note:** Both humans and animals will always select the easiest, smoothest route. That is to say if a track is partly covered at the edges with lots of stones, while the inner road is soft dirt—they will walk on the soft dirt (usually). Why? Because the stones will hurt their feet, and it is gentler to walk on a flat, soft, smooth surface.

From the footprint I can deduce that this is a male; weighing around 80 kilos (175 lbs) and is not carrying any rucksack or additional weight. His shoe size is 11 and he walks normally. Take a good look at the picture and try and deduce how I came to that conclusion.

The track surface is a mixture of hard rocky surface interspersed with soft sandy patches where it has been weathered. At this stage I will not bother looking for sign on the hard rock areas, but concentrate only on the soft sandy patches where sign will be clear.

Use logic to determine where you will find footprints, both animals and humans will walk the softest path, i.e, on the left where the soil is clear and not on the rough stones.

I move along at a good pace, constantly seeing the same footprint clearly in the sandy soil, but there is also something else that keeps cropping up: dog paw prints. Therefore, I am going to assume that the man I am following is with a dog.

Several times the same foot and paw print fall close together, then suddenly they do not. This tells me that the man has left the dog off the lead. Are there any human footprints to indicate where this happened, i.e, a set of prints close together.

The next thing I observe is that the track takes a full U-turn to the right and that both man and dog follow this route. This means that the man has no idea he is being followed and that he is not trying to throw me off the spoor by walking off the road. At this point, I consider if I could run ahead and try and catch up with my quarry. The

Dog feces are a clear indication of positive sign.

terrain indicates that this is not possible; therefore, I must continue to follow the spoor.

Some 50 meters further on, I find very fresh dog feces, which would indicate that my quarry was not far ahead. Then I heard the dog bark. This is a clear indication that the dog has smelt me, so I stand very still and do not move. While the wind in the right direction will blow your scent towards the dog, movement causes more scent to leave your body, if you do not move your scent picture is not so prominent.

In these early stages I want to concentrate on three factors: the footprint, confused sign, and the direction of travel. These three factors will help us start thinking and understanding the basics of tracking and help develop your skills. While doing this do not forget to look out for objects alien to the natural surround.

FOOTPRINTS

Much of visual tracking comes down to a footprint and while other signs are very important, without footprints we have very little evidence. The point is that humans stand still, walk, or run on their feet. It is the one point where we make contact with the earth. Given that most human footprints are embellished with some form of footwear, we have a fairly unique 'sign' that is visual and tractable. The same goes for animals and vehicles, as they all touch the earth in some form or another and most leave behind some definable 'sign' of their passage.

The human footprint is possibly the most important clue any tracker can locate; it is a definite, confirmed sign of man's presence. Let us look at it in detail.

The footprint is an impression in the earth. The degree of the impression will depend on many things. For example, did the human who made it step cleanly and evenly on the ground? Was the ground soft, hard, wet, or covered in snow?

These are the easy-to-see signs. Now let us see what else we can tell from a human footprint. For example:

- What size it is; if it is small and light it might indicate a woman.
- Is the tread clearly defined?
- Is there any disruption in the tread to indicate a damaged sole or trapped stone?
- Is the heel deeper than the toe which would indicate deception or someone walking backwards?
- How far apart is each step?
- If there are several clearly marked footprints, are they the same tread, as this might indicate military?
- If there are several types of print, this could mean a party of people.

So we see looking at a footprint provides us with a wide range of information and the professional tracker will glean as much intelligence as possible from the sign. Moreover, it will indicate to him the direction of travel which is an overall vital sign.

A Typical Footprint

This footprint was found during a walk through the forest. It has a distinctive pattern, more so because on the heel the design has one large block with no separation. At first glance we can see that it belongs to a male with roughly size 42 UK boots (11.5 US). The footprint is deep, which would indicate that the man is either overweight or carrying a load. The print is about two days old (it rained two days ago and this print was made after the rain). Strangely enough we saw no further footprints for over 1,000 meters, then suddenly in a soft spot was the same quarry, but this time the opposite foot—this gave us confirmation of his direction.

Confused Sign

In an area where there is little or no human footfall, any sign should be relatively easy to see and confirm. However, where an area has a heavy footfall, the sign, no matter how

The sign in this picture indicates at least two men, and a cross-country motorbike. The one print was made before the other as the tire of the bike has clipped the edge of the upper footprint. It's attention to this type of detail that makes a good tracker.

clear, can be extremely confused and in this case, a clear indication of print would be needed to actually ascertain you are following the right quarry. An example of this is where enemy soldiers have mingled with civilian refugees in order to confuse anyone tracking them.

You could pass several people over a large rocky dry area and they would leave little or no positive sign. Likewise, you could pass the same mixed group through a wet grass field in a farming community and whilst there is lots of sign, it would become confused. In all cases, you will need to hold the direction of travel.

DIRECTION OF TRAVEL (TRAVEL DIRECTION)

The direction of travel in tracking is the main direction your quarry is moving. Despite any tricks to throw the tracker of the scent, the quarry will have a goal or destination to which they intend to get to. Interpreting this end goal of the quarry by observing the signs left by the quarry determines the direction of travel. Once established, it is fairly simple to move at speed, checking periodically for confirmation sign.

Determining the direction of travel is critical in catching up with the quarry, enabling either the tracking to forge ahead or to instruct a blocking team to intercept the quarry at a given choke hole or known destination. This will require an interpretation of the terrain over which the quarry is moving and a best guess on their destination. If this cannot be defined, then a choke hold such as a valley between two high mountains or a shallow river crossing may prove successful in catching the quarry.

Cast

When you had a clear sign, but suddenly it has disappeared for more than say 50 meters, (especially where the ground is soft and shows other footprints), the quarry may have changed direction. This may be deliberate, or it could be a simple change of mind on behalf of the quarry. Was there another clear exit the quarry could have taken which was off the direction of travel? In such a case you should go back to the last known clear conclusive sign. From here you will need to walk round in a circle forward of the last sign.

The size of the cast will be down to the individual, but I always suggest around 30 meters at least.

As you walk, you should be looking for sign or any missed exit points; if the quarry has deliberately tried to mislead you then it will be difficult—if not, then you should be able to detect a new route. If this does not turn up any clue, then you have to try and eliminate the other possibilities.

Stand by the last known sign and look in the direction of travel the quarry was sticking to. Now look left and see what other natural exit points there are. Do the same on the right. Go to these and examine them in close detail. If you still find no sign, continue alone in the direction of march and hope to regain the sign. If nothing is found after 200 meters, then you have lost the quarry.

There will be many times when you lose the spoor alto-

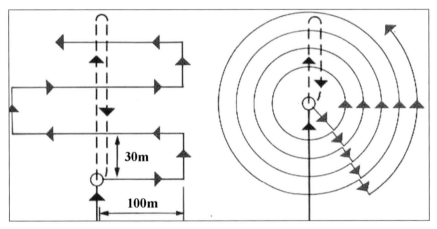

gether; don't worry, as this is fairly common. There are a number of search patterns which can be used to relocate spoor and allow you to continue. These include three methods:

Lost spoor is the bane of visual tracking, but it happens a lot; do not be disheartened for a lost spoor! Personally I have always found it best is to move forward from the last know conclusive sign.

1. A line search off the main direction of travel. This involves going back to the last known key sign. Mark this sign and move forward about 30 meters before moving out to the left and then out to the right, each time coming back to the line of travel. It is best if only the trackers do this, leaving the remainder of the team

at the marker. This will stop any misinterpretation or confusion.

2. In the 360-degree method, the tracker again starts at the last known key sign before making a circle which is ever increasing out to about 100 meters. Line of travel search involves using the natural direction of travel and pushing forward until fresh sign is found. In some cases trackers have moved several kilometers before re-engaging the spoor.

3. Your next move is to study your map and try and determine the direction the quarry may have taken. A good knowledge of map reading and visualization of the terrain ahead may well put you back on track.

OVERALL TRACKING OPERATION

Understanding the overall tracking operation is very important; why are you tracking the quarry—what is your endgame? Are you required to intercept the quarry or simply see where they are going?

Once the spoor is recognized and the tracker has established a general line of direction, it may not be necessary for him to continually look for sign. The natural lay of the terrain will indicate the direction of his quarry. For example, when we enter a house we logically go through the door. The same applies in the outdoors; if we are surrounded by trees and thick shrub with a natural opening roughly on our line of direction of travel, we will go through it. Occasionally the quarry will have climbed over a rotting log or crossed a clear soft patch of ground, leaving a 'confirmed sign' of his direction. More information can be gained if the tracker locates an overnight campsite or a rest area. By studying the campsite, he can establish vital information.

- Indentations or shelters will help indicate numbers.
- Camp fires and food scraps may determine the quarry's physical strength.
- Rubbish holes and make shift latrines will also reveal vital information.

DECEPTION

The quarry may well try a number of ploys to throw the tracker off the trail. Most efforts, unless properly executed, will be in vain and only confirm that the quarry is changing direction. Any delaying tactics the quarry may employ must buy more time in delaying the maneuver than time expended.

> **Author's Note:** This does not apply to military tracking where the enemy may leave behind mines or booby traps. While these are a clear indication of the enemy's presence, they have a tendency to slow the tracking team down considerably. One alternative is to leapfrog ahead in order to prevent running into booby traps and mines.

Walking Backwards

When a person walks backwards the length of his stride is shortened. The toe and ball of the foot will be more pronounced. Loose dirt, sand, or leaves will be dragged in the direction of the move. Always walk backwards on your heels lifting your knees with each step.

Brushing Track

This will only serve to *'sign post'* the intention of a possible change of direction to the tracker. Better to start side stepping over a distance, as this will reduce the spoor to nothing.

Crawling on Hands and Knees

The sudden absence of any top sign will indicate that the quarry is crawling or is injured. However, crawling is a good ploy to use if you come across a large animal trail.

Booby Traps

Unless the quarry has access to military mines and equipment, the time needed to construct a booby trap is time wasted . . . unless it is guaranteed to delay the tracker. If you do intend to build a booby trap, try laying three or four well concealed *'bluff'* trip wires, before the real thing. This will unease the tracker, making him slower and more cautious.

Booby traps are normally only associated with military tracking, and are designed to slow down the tracking team. Similarly they also slow up the quarry as they have to construct the booby trap.

Speed

If you are fit, speed may well put sufficient distance between yourself and the tracker, thus allowing the spoor to become cold. This is best achieved where the area is open and the ground sign you leave is light. If you intend to do this, attempt it in the early days of your escape.

Irrational Actions

Speed can be best applied with deceptions of irrational behavior. Doubling back parallel to the line of your march for some distance will throw the tracker. Climbing a rock face when it is not necessary, building a camp fire out in the open, will all confuse the tracker. Remember the tracker will not be alone. If you can put doubt of his abilities into the minds of others, it will help. Leaving a clear false trail will also work, but do not overdo it. Using railway lines for some distance—before jumping off—will grant you some time. Likewise, a large river can prove to be a simple and effortless mode of transport. Stealing a boat will bring attention to your method of escape—better to build a raft of driftwood and best to move at night.

Backtracking

By far the most effective method of deception is to use backtracking. This means establishing a direction of travel that is fairly clear to locate. Stick with this and then suddenly at a suitable location, move onto hard ground that will show no sign. Cut back parallel to your original direction of travel and then after several hundred yards, turn 90 degrees towards your desired final destination. If backtracking is done correctly it will throw off most professional trackers. The only drawback is if a tracker knows your final destination.

INFORMATION GAINED FROM SIGN

Having an understanding of the information from tracks and sign helps build up a picture of the quarry; it will also help you to get into the mind of those you are tracking. This information will help you determine if you should continue to track or if you should leapfrog ahead and intercept at a forward choke hole or prior to a possible final destination the quarry is headed for. Summarize in your mind and apply it to any other knowledge you might have of the quarry, and take into account all the information you have so far obtained:

- Age of track.
- Direction of movement.
- Number in the party.
- Speed of the movement.
- Confident or cautious movement.
- Weapons or equipment carried.
- Male or female.
- Morale of the group.
- Food eaten.
- Any deceptive measures used.

SECOND TRACKING EXERCISE

Now that the students have a basic idea of what to look for, I intend to make it slightly harder, but not so hard as to confuse and demoralize them. As with the first exercise, at each stage I want to explain to the students what they should be looking for and when they find it, what they are looking at and what information it provides.

First I will get two students to blaze a trail up to 1,000 meters long over soft, easy-to-spot earth, having a small stream somewhere along the trail will enhance the exercise. The idea is for the two students to make a series of signs so that the other members practice techniques of sign detection. You will need to do this exercise several times, rotating the students as you progress so everyone gets a go up front.

Those students blazing the track need to do a number of things so that the tracking party can identify the sign. These include:

- Dragging foliage behind.
- Walking in streams and shallow rivers.
- Walking backwards.
- Walking on rocks or hard surfaces.
- Walking backwards parallel to the original track.
- Jumping off the track.
- Laying false track.
- Stepping in another's prints to confuse numbers.
- Wearing rags or foliage on footwear.

As with the first exercise, you should instruct those students making the various signs to be generous in their markings, thus making it possible for those following to see and identify. The overall aim is to get the students looking for sign other than a footprint.

Always walk with the main party and once again remain towards the rear of the group so that the students do all the work. Understand the need for continued personal practice repeatedly stating the principles of tracking and explain the types of signs used.

Above all, get the students to fully comprehend the techniques for detecting sign and the information gained from sign. Before anyone can track you must know what you are looking for, and whether the sign is obvious, obscure, or hidden.

SUMMARY

Once the first two exercises have been carried out (several times for both), the students should have some idea of the skills necessary for visual tracking. It is now a matter of continuous practice and further development of these skills. While the individual can go for a walk in the forest and track another unknown walker, by far the best way to learn tracking is in a group.

For safety, always work in pairs and never alone. Make sure if you go into the wild you inform someone and if possible, have good communications. Decide before you go the type of exercise you intend to carry out. (For example: a simple clearly marker trail or one with some deception in it.) The idea is to learn and not to disappoint.

To assist you in creating your own tracking exercises, I have put several lesson plans describing how to do this at the end of the book. Enjoy the great outdoors!

Chapter 4

BORDER PATROL TRACKING

The complexities of border security are mainly dependent on the geography of the land mass to be protected. That is to say, a country surrounded by water such as Great Britain or Australia will concentrate its border protection on land and sea, although in the case of Great Britain, they also have the Channel Tunnel to deal with as this is a direct landlink to Europe. Countries such as America have two land borders; in the North with Canada and in the South with Mexico, with the other two sides facing the sea.

What this means is the resources and types of border patrol techniques used will vary accordingly. For example, those countries surrounded by sea do not have to worry too much about tunneling, or illegal immigrants climbing over fences, rather they will concentrate on unauthorized shipping and people arriving by air.

By contrast, a country which is surrounded by land on all sides will concentrate on monitoring its border on land. These borders may be a fence or a natural obstacle, such as a major river or jungle mountain range. In all cases, the relevant border protection agency will have to adapt its resources to the geography.

In many cases, it will not be just one agency that does all the protective work. For example, the United States Customs and Border Security are part of Homeland Security, which is a highly complex organization of many sub-agencies protecting its borders—these include the Coast Guard, airborne, and land based elements. Recently, America has redefined its border protection goal. The fact that resources are limited and the problems ever increasing, they have adopted a new strategy based on five objectives:

- Prevent terrorists and terrorist weapons from entering the United States.
- Effectively manage risk.
- Disrupt and dismantle transnational criminal organization (TCO's).
- Increase and sustain certainty of arrest; and
- Increase community engagement.

While the smuggling of drugs and illegal immigrants crossing the border remain a priority, today the major response of the border agencies is to prevent terrorists and terrorist weapons from entering the United States.

Among the countless types of border patrol professions, the Border Patrol Agent (CBP) is probably the most well-known. These agents are responsible for actively monitoring the Canadian and Mexican borders. Their primary role is to prevent terrorists, illegal immigrants, and contraband from entering the United States. One of the main duties of the Border Patrol Agent is the line watch, using covert surveillance methods to detect and prevent illegal border crossings. These agents cover a long stance of ground using cars, quads, and helicopters. They also have static positions which use a whole array of surveillance devices which will detect humans by both day and night.

Recently there has been an increase in the number of tunnels being constructed under the actual border in order to move in and out of the United States covertly. This is not a new idea and most likely derived from the publicity given to the Palestinians who have been tunneling into Israel for many years.

CBP Officers work at all ports of entry around the country, including international airports and seaports, as well as land border crossings. Their job might involve inspecting luggage for drugs, contraband, and other basic custom violations. In addition, there is also a staunch presence to prevent terrorism and the movement of weapons of mass destruction.

The work of the American Customs and Border Security is not dissimilar to the activities of Australia and the United Kingdom, other than the latter has a tendency to concentrate on their port and airport arrival facilities. Both also pay particular attention to any post arriving from overseas and it is their responsibility of help detect such items as *parcel and letter* bombs. In many cases it is the use of intelligence and computer based analysis that will tip off the agencies as to some unlawful activity, but the use of detector dogs also plays a vital part.

In all cases, it is the vigilance of the border protection agent that helps protect the people. They seem to have a highly developed sense of suspicion and are able to identify those who are not entering the country legally or are trying to defraud customs.

As we have already established, to enter a country there are limited options. You can arrive by air and take your chances with customs; you can arrive at a sea port and do

How strong the border security is will depend largely on terrain, where it is fairly flat and accessible, and then the stronger the security becomes.

the same. These are the normal routes the majority of people take; however, terrorists, drug cartels, criminals, and illegal immigrants can hardly turn up at JFK International Airport in New York with a suitcase full of cocaine . . . although a few have tried! These factions look for alternative routes to cross a border as a secretive way of eluding the border security forces.

In an effort to minimize the number of intrusions, the border protection agencies deploy a vast array of manpower and technical equipment. This ranges from tracking an illegal that has left obvious sign behind when penetrating the border to long range radar to intercept ships and aircraft. Despite all this effort, there have been some really ingenious ways of crossing international borders illegally, irrespective of the reason. In the main, this is achieved by stealth, using many techniques that have been perfected by the military. Covert infiltration can be done by land, sea, or air.

THE BORDER SECURITY AGENT

It is not my intention to elaborate on how borders are protected at sea or in the air, as this book is primarily about human tracking. Nor do I intend to repeat the techniques described in earlier chapters, as these will be common knowledge to any professional border security agent.

Depending on the length of the border patrolled, most agents will travel either by vehicle, quad or horseback. They will only go on foot when they have some evidence of an incursion or they are searching an area frequently used by illegal immigrants. In addition to their tracking skills, they will also be armed with a high-tech arsenal including seismic sensors, which detect people walking, heat-seeking cameras, lasers, and night vision goggles, which are used to pinpoint illegal immigrants crossing the border under the cover of darkness.

In general, the closer to the border fence the search starts, the better the chances of picking up some evidence of an illegal crossing. Disturbance on or around the fence itself such as bits of thread hanging from where the illegal has snagged their clothing, or heavy drop marks in the ground where they have jumped or dropped a heavy sack? Drug trafficking illegal's often carry heavy rucksacks which make deeper footprints.

Most patrols will check for sign by cutting a path in parallel to the fence, as agents look for disturbance in the dust, overturned rocks, crushed grass, and anthills that have been trodden on. Stepping on an anthill causes a high level of activity as the ants attempt to rebuild their mound; it can also determine how long ago the person passed by.

Direction is also easier to determine, as most illegal's walk in a straight line away from the fence. They will also have an objective, which might be heading for a small town or a highway where an associate will be waiting to pick them up. However, the border protection agents are wise to vehicles loitering around the border area and those heading away from it full of people.

In addition to searching near the border fence, most agencies will also have a second line of defense, whereby they use a blocking technique along known routes used by the illegal crossers. The local population often report sightings of people walking in remote areas or farmers close to the border who have had a theft from their property. This information all helps establish a line of travel taken by the illegal, so the agents can leapfrog ahead and intercept them.

However, by far the greatest and most obvious clue and the one that has led to the apprehension and prosecution of thousands of smugglers is the human footprint. For example, the agents working on the Southern border have a good idea of what is an American footprint and what is a Mexican footprint. Once several footprints have been detected and a line of travel established, the agents will inform the second line to intercept.

Whilst this sounds as if the border security agents are winning, they do not have it all their own way. To avoid detection, most border crossers will often wear some material strapped to the sole of their foot to obliterate their sign, with some even going as far as imitating a print that looks like cattle hooves. The agent's worst nightmare is when an illegal uses large booties made from blankets or rug material which barely leave any prints at all.

My aim now is to show the various methods of crossing a border illegally based on military knowledge and reports from border agencies around the world. As with all human tracking, we need a start start point, and to find that start point, we need to understand what is possible.

LAND

Land borders are normally fenced and patrolled, but no country can cover all of its borders, especially if they stretch

for any great distance over isolated country. Yet while an isolated place may offer easy access to cross a border, they are also desolate places and the chances of your survival are slim; the Skeleton Coasts of Namibia, or the barren coastline, of Eastern Oman are just two examples. Find yourself here and you will surely die or get discovered when calling for help.

If you try to get ashore in an isolated place using an aircraft or boat, then there is a very good chance that you will be detected, either on radar

Some borders, such as this isolated sea shore in Northern Malaysia, have become an escape haven for Muslim rebels in the south of Thailand.

or by one of the many patrols. Therefore, most people will seek to gain access by crossing a border in a simple place close to where they can blend into the local population. For example, when US CBP agents blocked off the routes in Arizona, the Mexicans simply switched to crossing the Rio Grande Valley of southern Texas. The cost of using helicopters, foot patrols, and trackers is extremely high, given that many of those that get caught and returned to Mexico will simply try again.

This is not just an American problem; some countries have major problems with terrorists using their borders in an effort to avoid capture in their own county. The south of

Thailand is a perfect example. Muslim terrorists regularly carry out bombings or shootings before slipping safely over the jungle border into Malaysia, where many of the population are Muslim and sympathetic.

METHODS OF INFILTRATION

There are numerous ways to cross a border without being detected; they take skill and a little daring, but if well-planned, can be executed successfully. While this may not look like anything to do with human tracking, it does provide an insight as the ingenuity of humans and this should help any tracker when it comes to looking for a start point. As we have already mentioned, one of the major advantages of any human tracker is to get into the head of the person he is tracking—learn how they think and what they might do.

Author's Note: There are few that would argue the fact that the British SAS is one of the best military units for getting into some places and getting out without being noticed. The American military have employed the SAS purely because their skills in infiltration and ex-filtration are legendry. It all started way back in WWII, when they worked hundreds of miles behind the German lines attacking airfield and fuel dumps. They continued honing and using their skills in almost every theatre of war they have ever fought in: Borneo, Oman, Northern Ireland, Iraq, and Afghanistan to name a few. While this book is primarily about tracking, it is always good to know some of the skills that are employed to pass over borders.

Many borders are protected by fences, but due to the length of the border it is difficult to stop infiltration; going under, over or through is fairly easy if you know how.

Fences

In many cases, fences form the basic border on land defense. Fences come in several different forms: some are single, some are double, and some are high or fitted with razor wire to prevent anyone climbing over. Conversely, vast distance with fencing is not only expensive, but needs maintenance and to be patrolled, because almost all fences can be penetrated.

To penetrate a fence you first need to study it first for its manufacturing construction, individual wire thickness, wire type, and wire pattern. You also need to look for any secondary problems, such as trip wires or hidden seismic sensors close to the fence. Check how the fence is actually set in the ground; the distance between supporting posts and what extra support is provided when the fence changes direction.

Construction is vital, for some manufacturing methods will allow for a certain number of links to be cut in order to collapse a large section of fence. The thickness of the wire is also important, especially if someone intends to cut the wire. Additionally, they may wish to climb the fence and, therefore, must be sure that it will bear their weight. The type of wire construction will influence the decision in how to tackle the fence. For example, if razor wire is evident, then some padding will be required if the intention is to climb the fence. In all, you have the choice of going over, through, or under. Here are a few Special Forces tips on how to tackle various types of fence:

Cutting

Most fences are constructed by weaving metal links together; cutting the links in a set pattern will reduce the number of cuts and shorten the escape time. On the other hand, solid mesh metal fences as used in some cases and these are best climbed using some home-made claw grip. This is easily constructed by heating a six inch nail and drilling it through a four inch length of a broom handle. When this is done and the nail is still warm, bend over the end two inches from the tip.

Climbing

Fences that can be climbed are often protected by a secondary barrier at the top—these can include razor wire, barbed wire, and rolling drums. In the case of the razor and barbed wire, these can normally be crossed by employing the 'Batman cloak.' Any thick matting, such as carpet or heavy canvas should be fashioned into a batman type cloak prior to climbing. The cloak will not normally get snagged as you climb and it is a simple matter of throwing it over your head and releasing it from your neck to achieve protection from the hazardous wire. Always climb where the fence is strongest, such as a retaining post or a corner. Be aware that many fences are alarmed and any attempt to climb them may trigger an alarm.

Going Under

If the fence looks too formidable, then you might consider going under it. If this is your chosen route, then you need to look for a spot that will facilitate a quick tunnel under the fence. This generally means looking for soft and loose soil or sandy conditions. In many cases the wire fence has been embedded directly into concrete. Do not be too deterred by this, as the concrete is rarely more than two feet deep. Do not start digging too close to the fence; about three feet back will provide a shallower trench for which to slide under. Make sure your trench is wide enough and deep enough for you to pass under quickly without getting caught up.

> **Author's Note:** No border is impenetrable; one of the most protected borders was the old East-West wall and fence which divided Germany after the war; however, this was crossed in so many ingenious ways. Several people built hot air balloons which lifted a small flat platform, barely large enough to fit four people. In September of 1979, Hans Strelczyk, a mechanic, and Gunter Wetzel, used their mechanical know-how to build a hot air balloon engine out of old propane cylinders. Their wives then pieced together a makeshift balloon from scraps of canvas and old bed sheets and the two couples, along with their four children, floated to freedom.

TUNNELS

Apart from simply crossing the border on foot and risking capture, many people have now taken to digging tunnels. The growth in digging tunnels under borders has increased during the last ten years and is now seen as a good way to move both people and contraband. The Israelites found one tunnel which had been dug from Gaza and

Palestinians smuggled goods into the Gaza Strip through tunnels under the Egypt-Gaza border at Rafah.

was over a mile long. The tunnel was lined with concrete and tall and wide enough for a person to walk or run though while carrying a heavy load. It also split and had several exits into Israel, some stretching over 600 yards into their territory.

More recently, the American border agents have discovered numerous tunnels, which when dug anywhere over the length of the US/Mexican border (3,145 km/1,954 mil) makes it extremely difficult to locate.

To counter this threat, border agents have resorted to using a series of modern technologies to detect the tunnels; these include:

Ground Penetrating Radar

This uses pulses of radio frequency energy to see beneath the surface. Whilst the technology has been around for over forty years, it is prone to false alarms and is limited in the best of soils to around 30 feet.

Electrical Resistivity

This relies on the fact that electricity cannot leap across an empty space, so when metal electrodes are placed in the ground they can detect a void. However, this is extremely expensive to implement as it would also show on the surface.

Seismic Waves

These cause waves to vibrate under the surface painting a picture of what is below. While still in its infancy, this method offers the best and most portable solution for detecting tunnels.

Human Observance

While this may be a basic way of detecting tunnels, it remains one of the most successful. When a tunnel is constructed or being constructed, there is normally a disturbance in the terrain, some small indication that a good human tracker will detect.

AIR

Crossing a border by air is extremely difficult, as the skies are monitored more than any other route. You will have normal air traffic control scanning the skies while controlling the massive amount of civilian aircraft moving around the world. You will have the military doing an over watch on the civilian controllers so your chances of 'sneaking in' are extremely limited . . . that said, borders have limitations. These limitations vary from country to country, but there is a general acceptance of some basic limits. For example, in the United States, the normal limit is 12nm (nautical miles), and 24nm when continuous; however, America also claims an economic limit of 200nm beyond that it is the open sea. Similar limitations apply to the air space.

The military have studied these limitations for many years, all with a view to covertly gaining illegal entry into a country so that their missions will go undetected. For example, it is possible for paratroopers to jump from an aircraft in one country and land in another, using the limitations imposed under normal law.

HAHO (High Altitude High Opening)

HAHO is a military method of air insertion where the parachutist exits the aircraft at a height of up to 30,000 feet and opens his parachute immediately. Using a RAM air

Borders have been crossed many times by the military; a High Altitude High Opening (HAHO) remains one of the best undetectable methods.

parachute, the parachutist can then glide for several miles; this allows them to infiltrate undetected across borders or major enemy concentrations. GPS can be used in flight to track the individual's position in relation to the earth's surface and the LZ area.

In the early 1980s the British SAS dropped a team of free-fallers off the south coast of England using the HAHO principle, all of which made it 10 miles into France (all with the approval of the French government).

WING SUITS

A wing suit provides extra surface area to the human body when in freefall. It was first used by a nineteen-year-old American from Los Angeles named Rex G. Finney in 1930. In 1997, a Bulgarian named Popov designed a more powerful suit creating lift which was able to slow the vertical speed to 18mph while gliding horizontally at speeds over 200mph. Then, by 2003, we saw the Austrian Felix Baumgartner jump from 9km (29,360 feet) up and successfully cross the English channel covering the 35km (21.8 miles) in just over fourteen minutes.

Wing suits have continued to improve and with the addition of small model aircraft jet packs, with even horizontal flight being achieved. Then on May 23, 2012, for the first time, a British stuntman, Gary Connery, jumped out of an aircraft and landed just using his wing suit. His landing runway was on a slope, which ended in a huge stack of cardboard boxes to break his fall. Gary crashed into the boxes and walked away unharmed.

A person in free-fall or wearing a wing suit is not likely to show up on any radar and the chance of a visual sighting is very remote. As the technology improves and the provision to reduce the landing speed to a safe level where it is possible to land without deploying a parachute, will be an ideal way of crossing any border.

Hot Air Balloon

As strange as it might sound, this form of transport is quite rare when it comes to crossing borders, yet it is, weather permitting, one of the best. Certainly if used at night, the chances of detection would be slim, as it is fairly silent and almost impossible to detect on radar. A single seated hot air balloon is generally known as a *'Cloud Hopper.'*

A cloud hopper is a single-person hot air balloon. There is no basket: the pilot wears a harness that is attached to the balloon, with a fuel tank and burner situated behind the pilot. Compared to conventional hot air balloons, cloud hoppers are very maneuverable and can land in relatively small areas; they are also very easy to store and transport between flights. In addition to enjoying the perfect, totally unobstructed view from thousands of feet up, the cloud hopper pilot can also fly at treetop level or low level walking through open fields in huge, forty-foot jumps. Cloud hoppers have been around for several decades, but are still fairly rare. However, they would prove an excellent way to cross any border and penetrate inland for many miles. If there are any drawbacks it is the illumination caused by the burner, which when lit can be seen for miles—the burner is required every so often in order to keep the balloon airborne.

Powered Para Glider

The powered Para glider has been around for many years - in addition it has been tested by the military as a vehicle to insert behind enemy lines, making it an ideal platform to skip over a border. Once again, it is highly unlikely it will show up on radar, but unlike the hot air balloon, it does make some noise, even when flying over 1,000 feet or so.

Author's Note: I was privileged to visit the USSCOM event in Tampa, Florida, earlier this year and was delighted to see the Stalker multi-purpose tactical vehicle on display. It is, without a doubt, an incredible machine capable of flying in the air or running as a normal vehicle on the ground. It has a pilot and two rear seats, which can also be converted into an evacuation stretcher, making it ideal for Special Forces. It took some ten years to develop, but it now has FAA approval and I am informed it is extremely easy to fly and almost crash proof.

Powered Para Gliders can easily cross borders, but nowadays they are often used by the border security services as they offer a swifter, and better view of any remote border areas.

The British SAS trialed a three-wheeled trike which had a parachute attached to it. Despite the powered Para glider offering great possibilities, the trial was abandoned due to the amount of accidents. The problem lay not with the Para glider, but with the would-be pilots, who in typical SAS fashion, hit the accelerator too hard on take-off, which placed the parachute too far behind them. They then immediately eased off which caused the opposite effect and so, caused the whole unit to obsolete back and forth in an extremely dangerous fashion—the project was abandoned.

That said, it is possible to see many civilian powered Para gliders floating around the skies and it would make an excellent method of crossing a border in areas where noise was not a concern.

SEA

The seas of the world are vast and they all have a coastline—some populated and some very remote—all ideal for making an illegal or covert entry. While the Coast Guard does a brilliant job of detecting and catching small ships, the one place that is fairly safe is under the water. That said, it is difficult to buy a submarine on eBay.

There are many ways of entering a country via the sea, but no matter what method is used, at some stage they must exit the sea and expose themselves. The one method used by Special Forces when approaching a border is to continue up a wide or remote river, taking them further inland.

Both surface and sub-surface methods have been adopted by various militaries around the world, including launching surfboards from a submarine, which the soldiers have had to hand paddle while lying flat on the board. Regardless of having to paddle some three miles, this proved an effective way of avoiding detection when penetrating the Fjords of Norway. Other methods include:

Divers

Most of us have seen the WWII movies where underwater divers have sneaked into the harbor of an enemy and planted mines on the sides of ships . . . well, little in practice has changed. True, the systems used have become *bubble less* and a lot more reliable, but their main aim is still to provide breathable air to a person underwater.

Almost all the Special Forces of the world have some unit which is dedicated to this role; in the US it is mainly the Navy SEALs, but there are also many other units, some of which are

Infiltration from the sea is also extremely popular with many illegal immigrants trying to gain access to another country. However, as this picture clearly illustrates, there are severe risks.

used to neutralize underwater mines, etc. The British have the Special Boat Section (SBS) which work alongside the SAS.

It is fairly obvious that anyone with these skills could easily gain access to a country and avoid detection; once again, the only drawback is that at some stage they have to exit the water on to land.

Klepper Canoe

The Klepper is a two-man collapsible canoe used by both the British SAS and SBS. This German designed canoe, which proved to be much lighter than the previously used Cockle II, came into service in the 1950s and remained until the mid 1980s. Despite its primitive design—the frame is made from hardwood Mountain Ash and Finnish Birch—the deck is covered with self-drying cotton woven with hemp and the hull material has a core of polyester cord, surrounded by rubber; ideal for clandestine insertions onto hostile coastlines. It can also be carried ashore and camouflaged by its crew. The canoe's skin is loose fitting with 'airsponsons' which run under each gunwale and are inflatable. It measures 5.2 meters long, 89cm wide and 61cm deep. It will pack into a bag 69cm x 58cm x 20cm.

SUMMARY

The purpose of border protection is to prevent illegal entry into a country. The purpose of a border protection officer is to identify and apprehend anyone attempting to do so. In order for them to do their job efficiently, they must be aware of all the possibilities open to those wishing to make an illegal entry.

While human tracking is only possible on land, no matter what method of entry is used, at some stage land must be touched and consequently a starting place for human tracking exists. Finding that starting point may prove extremely difficult, but it will be present. As with all tracking, knowledge of the unlawful deed and the methods they use is vital, getting inside the head of the enemy is critical to locating your start point.

In the face of all these ploys of deception mentioned in this chapter, the border protection agents do an outstanding job, their professionalism and local knowledge of the terrain they have to patrol is their mainstay in protecting their country from illegal entry.

Chapter 5

MILITARY TRACKING

Visual tracking in the modern military has almost diminished to the point where the skills have become lost. There are several reasons for this, one being technology the other being two World Wars. The First World War saw the military take to trenches and it was not too difficult to locate the enemy, as they were generally right in front of you. Troops left the trenches in the Second World War as mechanization took over and large units battled with tanks and armored cars. Tracking did not resurface as a military skill until the enemy and type of warfare changed. During the Rhodesian War, Selous Scouts were formed to track the bands of insurgents that regularly infiltrated their country. Likewise, in Borneo, the British sent small tracking parties over the border with Indonesia in order to intercept military patrols. The Vietnam War also saw the reemergence of visual trackers in a bid to locate the enemy.

While the bulk of the techniques required to carry out visual tracking are shown in Chapter Two, there are some aspects of military visual tracking that are unique. These fall mainly into two areas:

First, most enemies carry weapons.

Secondly, they may also ambush or lay booby traps for the tracker.

So the reader will forgive me if I do not repeat myself too much describing the 'sign' other than were it directly related to military visual tracking.

Additionally, military trackers will normally work as a team, and more often than not with a dog. There is a real threat when tracking an enemy and for this reason the tracker needs support; someone who can look out for him while he has his head down looking for sign. Therefore all the normal aspects and skills of a soldier become ever more important; the use of camouflage, concealment, field-craft, and good map reading will all aid the tracking team to be successful while keeping them safe.

In military terms, visual tracking is as important as personal skills, such as marksmanship and field craft. It is an integral part of the tracker's duty to use all their military skills (such as movement, light and noise discipline, concealment and deception) when tracking their quarry. While the main aim may be to track down the quarry, a soldier cannot forget that the enemy may lie in ambush or place obstacles in his path. Thus any military tracker must use all his military skills in order to follow, locate, and confront the enemy, all while maintaining his own concealment.

With military tracking comes a certain amount of danger, as in most cases the tracker will be following an armed enemy. At any stage the enemy may stop and prepare and ambush for the tracking party. They may also lay mines or IEDs in order to deter and slow anyone following them. Despite these hazards, many military units have success-fully located the enemy spoor and tracked them to their encampment.

There are also some advantages when tracking a military foe. For example, the tracking party will almost always have a clear start point. This might be the location of the enemy when it attacked an installation or an ambush point. Once the enemy has carried out its operation, they will normally move to a safe location as quickly as possible; if this is rebel-held territory, then the tracker will have a rough indication of the direction of travel. Additionally, the enemy will normally move as a group and, unless they are extremely well disci-plined, they will leave a lot of sign in their wake.

If the enemy is travelling a great distance that takes sev-eral days, it is almost certain that at some stage they will stop and rest. This rest may be for a few minutes or even overnight.

Once again, any place the enemy stops for a length of time will only serve to increase the amount of sign left behind.

So despite the dangers of military tracking, there are some advantages; neverthe-less, any tracking party will need to be extremely alert to the possibility of ambush and booby traps. In many military operations, soldiers that are tracking the enemy become alert for signs of enemy activ-ity. These signs help the sol-dier become aware of the ene-my's presence and gives them time to react. In countless

Running into an ambush is one of the major dangers of military tracking.

cases it is a matter of pure instinct and the reliance of human senses that have long been ignored. Any professional military tracker will tell you that at certain moments during the follow they experience a gut feeling—usually just seconds before all hell breaks loose.

While there have been many excellent reports of successful tracking over the past half century, several units stand out above the rest. These include the Selous Scouts of Rhodesia, the British SAS operating during the Malayan campaign, the Australian SAS, and the Americans in Vietnam.

Author's Note: While I am aware that many of the American military have carried out successful tracking operations in various counties, I do not know the details and thus cannot comment on them. The reader will forgive my limited knowledge in this case.

SELOUS SCOUTS

During the war in Rhodesia from 1966 to 1980, C Squadron (there is no C Squadron in the British SAS—this slot was taken up by the Rhodesians), was commanded for much of that time by Major Brian Robinson, and played a major role in mounting cross-border operations in Zambia and Mozambique. They inflicted heavy losses on ZIPRA and ZANLA forces and by June 1978, having expanded to three squadrons under command of Lieutenant Colonel Garth Barrett, the unit was re-designated 1st SAS Regiment. When they were finally disbanded, most of the 1st SAS Regiment had left Africa and joined the British SAS.

SAS BORNEO

In the early 1960s, Malaya sought to bring together Singapore, Sabah, Sarawak, Brunei, and itself under the Federation of Malaya, an idea fully supported by Britain. The Indonesian President Sukarno, however, vehemently opposed the move as it threatened his own designs on the rest of Borneo. The first sign of trouble came in 1962 in the Sultanate of Brunei, when a small Indonesian-backed anti-Malaysian element rebelled. The British forces, however, quickly ended this revolt.

By early 1963, the situation had worsened, with well-trained Indonesian insurgents infiltrating over the border from the Kalimantan region. In response, Britain raised a force of Malaysian, Commonwealth, and British troops—including the SAS—to deal with the situation.

As well as patrolling the border, the SAS took on another very important task—that of winning the 'hearts and minds' of the native people. By gaining an understanding of their lifestyle and language, by living with them and dispensing medical aid when necessary, the SAS gained important allies in intelligence gathering. The local people, who still crossed the border freely into Kalimantan to trade their goods, often brought back valuable information on Indonesian troop movements. The SAS recruited some of these local people, who were excellent trackers and trained them as Border Scouts.

Author's Note: Captain Malcolm McGillivray had command of 2 Troop during the emergency in Borneo. His mission was to search Mount Kalimantan for Indonesian units and destroy them. His team consisted of ten SAS, including Corporal Roberts, Troopers Franks, Henry and Bilbao, and Condie, Callan, and Shipley, as well as twenty-one scouts, gathered from local tribesmen.

McGillivray and his men set out on July 9, 1965, crossing the River Bemban and trekking up through the foothills. This first part of the journey took five days. On the sixth day, McGillivray split the men into six patrols to search the area for any signs of enemy occupation. At first, the intelligence, which they had been working on, seemed to be right on target as one of the patrols returned with the news that they had found boot prints and smelled cooking. No matter how hard they searched, there was no enemy camp to be found. However, they did relocate the track which was heavy with enemy sign and estimated that it must be a supply route to Batu Hitam. After a reconnaissance, McGillivray decided that the best ambush point was at a gully just northwest of the river. Here, the gully was spanned by a large log. Anyone trapped on the log when the shooting began did not stand much of a chance.

McGillivray took Slater and two others with him to keep a watch on this point. A group of locals appeared, carrying heavy bundles and crossed over the gully using the log. Convinced now that this was definitely the place for the ambush, McGillivray sent the two men back to round up the rest of the team while he stayed at the point with Slater. Soon after, a group of armed enemy soldiers, about fifteen in number came in view.

However, at such close range and with such odds, it was deemed unwise to take them on without the rest of his men present.

The rest of McGillivray's men arrived and he selected fifteen men, including Corporal Roberts, to form the first watch. This group was divided into three separate sections, each with a different responsibility during any attack. The men were all then placed individually in one long line, five to twenty yards away from the track. The 'killer' team, consisting of Roberts, Franks, Nibau, and three scouts were placed alongside the log. The group on the right-hand side, consisting of Condie, Callan, Shipley, and three scouts were responsible for keeping a watch for the enemy approaching and informing the others by means of pulling a cord that ran down the line. They were also charged with stopping any escapees and warning of the likelihood of a counter-attack. The left-hand side group assumed a similar role to the right.

An hour later it began to rain heavily, with the storm creating difficult conditions for the watchers. Then, without warning, halfway through the afternoon, the soldiers that had passed through before, returned. They were obviously wet and uncomfortable as they traveled at a very quick pace. Nevertheless they were spotted by Shipley, who was sitting on the right-hand flank and gave five tugs on the cord. Roberts alerted his men just as five of the soldiers stepped onto the log. The 'killer' group all took aim and fired, sending the Indonesians tumbling into the gully below. Roberts, thinking that all had been killed, gave the order to stop firing. Bilbao, however, spotted one of them trying to crawl away and let off two more rounds to stop him, as he had been previously instructed.

However, the five men on the log had only been one part of the returning soldiers, as the men on the right flank were about to find out. Alerted by the gunfire, a large enemy group disappeared into cover next to the track and let loose with their rifles and machine guns towards the ambushers. It looked as though they were about to storm the position from the right flank. The thick jungle cover slowed the battle and although the gunfire continued from the enemy's position, they did not seem as willing to advance as they had been before. The bodies of two Indonesian soldiers lay under a log close to Roberts' position, but as yet they had not been searched for weapons or papers. McGillivray turned to two local scouts nearby and asked them to carry out the task, something which they seemed to be only too glad to do. As the SAS group gathered together to prepare for the journey out, the scouts that McGillivray had sent to search the bodies returned, not only with enemy weapons and papers, but also their heads.

Although revolted by the practice, McGillivray and his men made no attempt to stop it or dampen the tribesmen's celebration in any way. It seemed an important thing to the Iban, a business which appeared to raise their morale and as McGillivray and his men readily recognized, if it raised the Iban's morale, such an act was likely to lower the Indonesians once they found the headless bodies of their comrades in the jungle. Nevertheless, the head-hunting incident would be wonderful news for any hostile propagandists if it got out so it was decided that the whole event needed to be kept quiet. And despite rumors, during his future years in the Regiment, McGillivray always denied that he had witnessed any head-hunting.

AUSTRALIANS IN VIETNAM

Australia was involved in the Vietnam War from 1962 until 1973, and troops were deployed again in April 1975 during the fall of Saigon. The initial commitment was only thirty men, but at its peak there were over 7,672 Australian service personnel in the province. In total, some 60,000 Australians served in the War with 521 killed and another 3,000 wounded.

The Australian SAS, together with troops from New Zealand's SAS, were deployed to provide intelligence to the 1st Australian Task Force (1ATF) and US forces deployed in the region.

Author's Note: On June 21, 1967, during the conflict in Vietnam, an SASR patrol was inserted at 18:05 hours into an area five kilometers northeast of Thua Tich in the Phuoc Tuy Province. The patrol, part of 1 Squadron, was commanded by Sergeant Barry Glover, an excellent soldier and great tracker.

The drop at the LZ was uneventful and the patrol moved forward into the thick undergrowth. However, a short way from the LZ, the patrol came across a Vietcong (VC) sign which led them directly to a VC camp from which they heard voices. As they had not been compromised, Tex Glover decided to swing around the VC camp and move north. Once at a safe distance and having established a lying-up position (LUP), they would be able to radio HQ for back-up and wipe out the camp. They had not gone far when suddenly three VC appeared out of the jungle, causing one of the patrolmen to open fire, killing two of them. Knowing that their shots would have been heard, the patrol quickly moved out from the position. The patrol tried to distance itself, but it wasn't long before the rear man, Corporal Steve Bloomfield, reported that the enemy was getting closer.

Glover decided that if they kept retreating, the enemy would soon have the advantage. Placing his men into a temporary defensive position, they waited in ambush for the VC to advance . . . they didn't have long to wait. About twenty VC approached the patrol's position and a fierce firefight began. During the battle, some of the VC were killed and the patrol managed to disengage and move off. Unseen, they managed to make it to the LUP. The VC did not give up easily, however, and mortared the area throughout the night. The bombardment was so relentless that Glover was not able to request an extraction until midday on June 22.

The helicopters that arrived to take them out came under heavy fire from VC positions. Even the suppressive fire from the helicopter gunships could not silence them. Despite the danger, the helicopters stayed on site, dropping down lines in order to winch the patrol members up through the jungle canopy; a height of some 129 feet. Those patrol members still on the ground provided covering fire until everyone was safely extracted. Once aboard the helicopter, the men were able to add their weapons to the aircraft's own, raining down a hail of bullets on the VC. Amazingly, both patrol and rescue helicopters emerged from the battle unscathed and returned to the 1 Squadron base at Nui Dat. It was later estimated that Tex Glover's patrol had faced a VC force of company strength.

AMERICA IN VIETNAM

In 1966, with the War in Vietnam escalating, General Westmoreland, who was impressed by the UK's use of trackers in fighting Communist insurgents in Malaysia, sent Lieutenant Colonel Starry along with a small team to the British Jungle Warfare School in Malaysia. This resulted in American soldiers being trained in tracking techniques by British and New Zealand instructors. Some 140 soldiers were initially sent to the British Jungle Training School in Malaysia.

After the soldiers completed their Tracker Training they returned to Vietnam. The trackers were organized into four teams per division and broken down further into two elements per brigade. The mission of the Combat Tracking Teams (CTT) was to discreetly make contact with the enemy within their specific area of operations. The CTT were usually worked ahead of a supporting unit by a platoon of regular soldiers in order to maintain noise discipline and element of surprise. The first units were assigned to the 25th Infantry Division in February of 1967 and operated very successfully until the end of the War.

Author's Note: I am pleased to say that I contacted some of the former members of the original American tracking teams and I am grateful for their support. I have no actual stories of the American tracking teams, other than those posted on the Internet. What I do know is that a lot of good Americans volunteered for these teams and a lot died in Vietnam.

Last year, I was in Washington and decided to pay my respects to all those who'd lost their lives during the War in Vietnam. For those who have never seen the 'Wall,' you should go—the artist did a wonderful job as every name is etched on a black stone wall and when you place your hand upon the wall you can practically feel the souls of the 58,195 who gave their lives.

US trackers with a dog during the Vietnam War.

THE ENEMY

If your tracking operation involves hunting down members of a terrorist group, do as much research on their organization as possible. In many cases, terrorists are given much more credit than they deserve. Do not accept that the enemy is a real soldier or an excellent bushman even when they are indigenous to that continent. In Rhodesia, Angola, and Mozambique, terrorists have been found wandering, lost and starving, because they were city or town born and had no wilderness survival skills whatsoever.

Author's Note: Over the years, I have fought in many small wars and the enemy has always been different. In some cases they were very professional, with excellent military skills and knowledge of tactics; these, however, were few and far between. Most terrorists are little more than thugs with guns. Researching your enemy is a real must for any visual tracker, add to this any personal contact you have with the enemy and making your assessment accordingly. It has been my personal belief that when a professional soldier stands his ground—or even advances forward—no matter how good the enemy, they will retreat.

The better you know your enemy, their causes, objectives, personalities, practices, and tactics, the more your tactical edge over them. Once you understand the enemy, your task will become much easier. Watch their standard operating procedures (SOPs) and record their responses to your tactics: absorb every piece of information available to you.

LOCAL TRACKERS

It may be possible to integrate yourself within the local population, but do so with great caution . . . but make sure to keep in mind that local trackers will be of great advantage to any tracker team, as they will know their own backyard in far greater detail than you.

When seeking to employ native trackers, it is best to inquire with the local headman or chief and deal through him directly. Not only will they know who the most reliable local trackers are, but they should also be able to provide insight to the local politics. Talk frankly, and be friendly but firm. Offer assistance in the form of medical aid and do not offer money in the initial stages. Learn how the local people view the enemy; they may be in sympathy with the rebel cause or scared by intimidation. Alleviate their fears if possible and be prepared to offer top wages and protection for the trackers and their families. If the enemy has already visited the village, check out any native male of military age who might be seeking revenge against the rebels and use this to your benefit. Never promise them anything unless you can deliver immediately and never lie to them. In many situations only mutual trust and respect will keep them loyal to you.

Once you have employed your local trackers, keep a record of how they perform. Good trackers will be able to tell how far ahead the enemy is and allow you to put blocking troops in position. If you come under enemy fire, make sure you protect them—but never arm them unless you are 100 percent sure of their loyalty. Never discard any intelligence you get from the local populous; they know the tracks through the jungle or bush and will instantly recognize if any terrorist group has passed through the area. Farmers in Northern Ireland often reported the discovery of an electrical wire which ran across their fields. Further investigation

revealed a bomb on one end and a firing point on the other.

If your team does employ local trackers, then learn from them, ask questions as to why they are going a particular way, and watch how they move and observe.

Local Trackers are far superior at visual tracking especially when working in their own backyard.

The British SAS had a firm policy of making friends with the locals by winning their hearts and minds and were able to gain vital intelligence about any enemy activity in the area.

Author's Note: The British SAS made a habit of conducting 'hearts and minds' mainly by providing medical aid to the local villages. They would always approach the headman or village elders first on every visit. Once it was known that the SAS were hunting Indonesian soldiers, many of the local young men eagerly jumped at the opportunity to act as local scouts.

TRACKING TEAMS

While a tracking team can be of any size, it is generally formed around the structure of a team leader and several good trackers with the same amount of cover men for security and a radio operator. When moving forward, the best tracker will take point with the other trackers covering the flanks. The team leader will stay in the middle with his radio operator immediately behind him. In general, most of the team could cross skill, as many were excellent trackers. The role is to be dictated by the unit commander who will either plan an operation to locate an enemy camp or do an immediate follow-up after any enemy engagement. The team will be required to carry out the following type of tasks:

- Provide tracker skills to a larger follow-up operation.
- Locate the start point spoor after an engagement.
- Track and terminate a small group of insurgents.
- Locate enemy camps.

When friendly troops have been ambushed, they are used to track down the enemy ambush party, or vice versa when enemy soldiers have survived an ambush. Enemy units that continuously fire mortars or rockets at a friendly base will need to be located. Tracking teams may also be sent out in anticipation of an enemy build-up that may come in the form of an all-out attack. They are also used to locate downed pilots or missing personnel after a fire-fight.

Depending on the terrain, the team would move along the direction of travel in the following combinations. The best tracker would always lead once the spoor has been located. A minimum of two flankers—who are also good trackers—would move out to the sides (terrain permitting). Somewhere in the center of this group—but staying clear of the quarry spoor—would be the tracker control who would keep an eye on their location and report back to base on progress. This would normally be a team commander (tracker trained) and a signaler. The team would move forward as quickly as possible developing a tracking plan as they moved:

- Search and relocate the spoor quickly if it had been lost.
- Marking the last clear sign for reference when the spoor is lost.
- Automatically adopt a standard search pattern when looking for lost spoor.
- Note when the spoor hade deviated a lot from the direction of travel.
- Automatically rotate duties as the lead tracker becomes fatigued.
- Input all sign and observations from all members to form a cohesive tracking picture.

As with any unit, the terrain and dangers will dictate team formations. In fairly open ground the team will expand, with the flankers moving slightly ahead of the lead tracker who will be in direct contact with the spoor. Should the spoor suddenly move off to the left or right then the flankers should pick this up. When spoor is lost the flank trackers will circle inwards towards the lead tracker in the hope of quickly regaining it. While they are doing this the lead tracker will normally carry out a smaller 10 meter, 360 degree search. In most cases this is sufficient to relocate the spoor. If not, then a major box search will be required.

Author's Note: This can be a worrying procedure as both flankers and lead trackers will be head down looking for the lost spoor. That is the reason why flankers work in pairs, so that the second man can provide heads-up cover in the event of an enemy trap. The team leader should remain stationary to provide protection and fire support should it be a trap. This basic procedure should be carried out in both close country and thick bush.

In thick, close bush or secondary jungle the team will normally move in single file. This is simply dictated by the fact that the noise and difficulty of moving in any form of extended line would prove dangerous and difficult. As a result, the lead tracker will normally push his way up front by at least

10 meters while the team commander and signaler are followed by the flankers. Upon lost spoor the lead tracker will do the same as before and carry out a small 360 degree search from his direction of travel. If the spoor is still not located then the flankers will carry out a larger 360 degree sweep stopping when they are level with the lead tracker.

Warning: Getting separated in thick bush is not a good idea and the team commander should be aware of the position and route taken by every team member. If an ambush is sprung while the team is doing a wide-area search in thick bush their chances of survival are greatly reduced.

Tracking drills require that everyone take part, it is the only way a tracking team can survive and usefully complete their mission.

TRACKING DRILL

Because of the dangers of military tracking, there have to be set drills which will reduce the possibility of the enemy ambushing or booby trapping the path. The visual tracker must be fully aware of the possibilities of enemy activity in the area they are operating in. The type of terrain over which you are tracking will dictate your speed and observation skills. For example, it is possible to look forward a greater distance in the desert than it is in the jungle.

- Rehearse your team in walking silently, check for any metallic items that may cause sound.
- Make sure that each team member knows their position and practice anti-ambush drills.

- Keep the radio switched off until required.
- Make a firm assessment of the general direction.
- Look forward along the direction of travel, keep your head still and only let your eye sweep left and right looking through the vegetation or over the desert floor.
- Visually connect the track with your eyes by comparing age, eliminate the older track.
- Check both left and right as you move forward looking for any sign of deception.
- Memorize the footprint, keep this profile as your *'key'* for the whole track.
- Be alert to all sounds, smells, movement and vegetation in an unnatural state.
- Move forward, not as a tracker, but as a lead scout. Remember, you are most vulnerable when moving or passing through a funneled point.
- Eliminate opening by comparing age, be aware of over-deliberate sign, as this may lure you into a booby trap.
- Take care when moving through noisy vegetation or over rocky ground which causes stone slip noise.

Halts

From time to time, tracking units will need to stop and take a rest or make a decision on what to do next. No matter what the halt is called for, every member of the team must obey the basic military strategy of adopting all-around defense. For very short halts where the lead tracker is trying to define the route forward or where an obstacle has presented itself, everyone should simply drop to one knee and face their arch of observation and field of fire.

For long halts such as a physical rest, then all members should lie down, but again retain observation over their own field of fire. Where cooking breaks are allowed over very long tracks, then the team should work in pairs—all-around defense is critical to the survival of any military unit. If the halt is due to a nighttime stop, then normal perimeter defenses should be set up. This may entail the planting of early warning tripwires and *'Claymore'* mines etc. Any nighttime halt should take place when there is sufficient light to lay down any perimeter security. Ideally, it is best to rest and eat in one location then move on to another before bedding down for the night. Environment will dictate the need for sentries. For example, if you are in the desert or an environment where people can still move around at night, make sure to post sentries. If you are deep in the jungle, sentries are not required, as it is almost impossible to navigate or even walk through thick jungle during the hours of darkness.

Author's Note: This does not apply when the tracking patrol is close to a village or an inter-connecting jungle track that is well worn. It only applies if the patrol is deep into secondary jungle.

Obstacle crossing in itself is dangerous, when tracking a quarry it becomes a point of possible ambush as well, because you are vulnerable.

OBSTACLE CROSSING

If the halt occurs due to an obstacle, such as a river, then the team should work out a procedure that will get them across securely. Major obstacles such as rivers or major track junctions are ideal places for the enemy to lay in ambush. Likewise, a river or major track junction is also a source of key sign for any tracker. While the obstacle needs to be crossed securely, it must be done in such a way as to allow the lead tracker first view of the river banks or track so he can ascertain any direction change of the quarry.

Normal infantry tactics do not apply when a tracking team reaches a major obstacle such as rivers, roads, or trails. The rest of the team will remain to provide cover while the lead tracker goes ahead and crosses the obstacle first. This is done to preserve the spoor and stop any unnecessary corruption. Once the tracker is safely across and has relocated the spoor, the rest of the team will follow using normal covering tactics. In all cases, the priority must remain focused on maintaining visual contact with the spoor.

TRACKING REPORT

It is vital that the tracking team keep the overall commander in the picture as to what is happening; this is normally done in the form of a quick 'sitrep' (situation report) or a detailed tracking report. This job normally falls to the team signaler, and depending on the type of communications device, will normally take place when the tracking team commander calls a brief halt. A sample of a tracking report might follow the following format.

The report will normally start with the tracking team's location, using standard military grid system, i.e, a six figure location might be 348953. This will put the team within a 100 meter radius.

'D' will denote the direction of travel.

'N' can be used to indicate how many troops you are following in your estimation.

'A' is the age of spoor as accurately as possible.

'T' is sometime used to indicate the type of quarry being followed, i.e, enemy boots or bare foot locals.

From these reports the operational commander can adjust his plans and organize any follow-up troops to intercede the enemy. In addition, the operational commander may be in possession of information from other tracking units allowing him to make better decisions.

Follow-Up Troops

Follow-up troops can be either in close proximity to the trackers or remain in a base camp where they can be deployed by vehicles or airborne methods. Follow-up troops are normally deployed at the discretion of the overall operational commander depending on the size and location of the enemy. If the tracking team has entered a foreign country in hot pursuit, all the political ramifications need be taken into account before follow-up troops can be deployed. In many cases the follow-up troops will also act in the search and rescue role, should the tracing team become lost, injured or caught in an ambush.

Support troops used for blocking the quarry are normally placed way out in front of the direction of travel; a parachute deployment is quick and effective.

USE OF AIRCRAFT FOR TRACKING

Light aircraft, parachute planes, and helicopters can provide rapid deployment of support troops. These may be blocking troops landed in front of the enemy or support when the tracking team has been drawn into an ambush. The one drawback with aircrafts of any type is the noise which can act as a warning to any enemy group being tracked. However, in most wars, there is now so much airborne activity that the odd extra one draws little or no attention.

OVERNIGHT CAMP

There may well be many an occasion when the tracking team is forced to spend the night out on the trail. Once again this is dependent on the environment and the quarry. If the quarry is moving through thick, dense jungle they too will be forced to stop as soon as it becomes dark. Similarly, a fatigued quarry will also need to rest and the best time is during the hours of darkness. The tracking team commander must make an evaluated guess into what the conditions of travel are for both the team and the quarry. Likewise, both the team and the quarry will have a limited endurance which, if ignored, can lead them to become complacent and make mistakes.

Should you find a resting place, you should count the number of footprints and any sleeping locations. The number of weapon marks will all help build a clear picture of how many of the enemy you are following and how well armed they are.

Coming across an enemy overnight rest area the following day, it will be easy to determine how far ahead the enemy is. Given that most troops will break camp around first light, it is simply a matter of noting the time of camp discovery. It is then up to the tracking team leader to make the decision as to push ahead or remain and glean as much information as possible from the enemy camp area.

Any overnight stop should be sufficiently off the direction of travel as to limit the amount of noise or light the tracking team may produce during their rest period.

ANTI-AMBUSH DRILLS

Anti-ambush drills vary slightly, depending on the terrain and the type and strength of the patrol. If, for example, we have a tracking team operating in the jungle, the chances are it will not exceed approximately eight soldiers. The difference from a normal patrol is the lead scout is also the senior tracker. This means they have to track and be alert to an ambush or booby trap.

Moving through jungle is generally done in single file due to the narrow tracks, which means, with good spacing, the team may be spread out over 30 meters or so. Control over the team being driven by the lead scout/tracker, or in some instances the team commander. The type of control used is hand signals, most of which we have all seen in the movies. For example, if the lead scout holds up his hand palm facing forward it means stop—and everyone in the patrol will hold their current position. Waving his hand forward will mean continue. Doing the same movement, but this time clenching his fist, will mean the lead scout has seen or heard something and requires everyone to hold position, squat down, and be alert. There are lots of different hand signals and some change army to army, but for tracking there are a few that indicate basic meanings.

A tracking team can effectively use hand signals instead of talking to indicate their actions.

- Hand up palm forward: halt
- Hand down palm waving forward: advance
- Hand up clenched fist: freeze / enemy / seen or heard something
- Hand up clenched fist pumped up and down: move rapidly
- Hand formed as if holding a telescope and held to the eye: looking or going for a look
- Single finger held to the lips: silent

- Two fingers touching shoulder: get commander up to me
- Hand to the ear: listen

While this may look efficient and hand signals do work extremely well, it is my firm belief that the first sign of any enemy ambush is when they open fire, or the lead scout who has spotted them will bring up his weapon to fire.

Having been in three ambushes during my military service, each time it was the experience of the lead scout that triggered our actions, having seen the enemy first. It is fair to say that the British SAS very rarely move along jungle tracks; instead they prefer to move directly through the jungle bush often contouring the ground. This gives them the advantage of surprise as most normal infantry will move along the easy paths, i.e, jungle tracks. Therefore my advice is simple:

Anti-ambush drills require you to move out of the 'killing zone' as fast as your legs will carry you.

- If the lead scout raises his weapon, be prepared for an immediate fire-fight and drop to the ground.
- Patrol members in file should work in pairs, being prepared to lay down suppressive fire the moment the lead scout either indicates enemy to his front or he actually opens fire. In the case where the lead scout has seen the enemy but the enemy has not seen the patrol, then the commander may order an immediate ambush or withdrawal, depending on estimated enemy force size.
- In the case that the lead scout opens fire, the general drill is for everyone to fall down facing the direction of the fire (the exception is tail end Charlie who will look to and protect the rear). A well-practiced format of falling left and right off the center line is best, as this will allow each pair in turn to provide covering fire as the team lead scout first peel back down the way they came. The whole team will continue this movement until the team commander screams *on me* indicating a rally point, at which stage initial contact should have been broken and the team commander has found a hole for everyone to jump

into. A quick head check, and if everyone is okay, then the patrol will move off into thick jungle.

- In the event that the lead scout has been shot, then it is up to the first forward pair to go get him and drag him out of the kill zone.
- If this drill is exercised and rehearsed prior to leaving the base, then there is a good chance that the patrol will successfully extract itself in good order.
- In the event where the lead scout goes down in a hail of fire and can positively be identified as dead, then the patrol should extract in the same way, leaving the dead behind.

Here are a few tips to avoid an ambush:

- In a dangerous situation, be aggressive—no doubt the enemy will be—let them know you are not to be messed with, so get your shot in first.
- Know your own team's strengths and weaknesses.
- Know your territory and its inhabitants.
- Better to be known than be a stranger to the area and its inhabitants.
- Know when to stop tracking and get the hell out—always have an escape route planned.

Many military units will say 'we leave no one behind,' but it has always been my personal opinion that there is no point in risking more lives for someone who is confirmed dead or in a position where any rescue attempt will only jeopardize the whole team. That said, if there is even a slender hope of rescuing a comrade, it should be done, as most soldiers will tell you it all depends on the circumstances and a logical decision being made at the time.

However, contacts in close jungle are not cut and dry; not knowing the enemy's strength or their reaction may provide an opportunity to revisit the ambush site. The enemy returning fire is a good indication of their size and if it fades, it may well mean they are retreating; the decision is then down to the patrol commander as to what they do next. This decision will normally be taken at the rally point where everyone can have their input.

It is not my intention to delve too heavily into ambush techniques, as in most tracking environments the enemy will only consider ambushing or laying a booby trap if they are certain they are being followed.

There are two things against laying an ambush against a tracking team following you:

Firstly, the enemy must wait for the tracking party to catch up, thus annulling any lead they may have established. Booby traps take time to prepare if they are to work correctly and there is no guarantee that the tracking party will spot an unmanned booby trap. All booby traps should be well recorded if left unattended as they can often kill or wound your own forces, or innocent civilians.

My personal experience has shown that if an ambush is set up, it is due to the fact that there is clear evidence of the enemy presence and a good spot can be chosen for the ambush. The second case is where a tracking patrol has become aware of an enemy approach and has acted accordingly. Booby traps are a good idea if you want to catch out the enemy using a well-known supply route or a patrol wishes to cover their rear. The British SAS used what is known as an 'A-type' ambush.

A-Type Ambush

An A-Type ambush is a series of unmanned explosive devices that are set up and left for the enemy to walk into. The explosive could vary from WP (white phosphorus), to grenades and Claymore mines, all linked by a detonating cord into a single triggering device. They were mainly used on set routes by the enemy for ferrying weapons. A-Type ambushes were laid during Borneo operations and during the Oman War. They were always clearly recorded and removed if not triggered.

Author's Note: During the Oman War of 1970–75, the British SAS established a series of firm locations along the lip of the Jebel mountain. Previously, the enemy had used this high elevation to rain down mortars and rockets on the airfield near the coastal town of Salala. The establishment of the firm bases, known as Diana's, meant that the enemy's weapons were now out of range. Conversely, they now trained their weapons on the fixed locations. Surprisingly enough, they would almost always fire from roughly the same location, which was not difficult to track down. Once we had located the exact spot (which indicated the position of a mortar base-plate), we set to mining the area with some rather nasty American-made jumping mines. Myself and a close friend, *'Geordie'* Barker (who died in Oman—God rest his soul), were given the task of laying the mines. Some days later we heard the mines explode and, upon further investigation, we discovered we had killed three of the enemy (there were only two bodies, but there were five arms so we claimed three kills).

MILITARY TECHNOLOGY

Sensors, cameras, and Unmanned Aerial Vehicles (UAV) have become very popular to detect human presence and passage. These types of technologies, although impressive, have been used to replace the human on the ground instead of augmenting them. These costly technologies have been over relied upon and have not proven their worth over the long run. The high incidence of illegal activity that crosses the United States border every day is a good example of that. Unattended Ground Sensors (UGS) utilized in both military and law enforcement sectors have never really proven their worth. Insurgents, as well as criminals, have always been able to find and render them inoperable. Cameras employed in static positions or mounted on UAV platforms, although used widely, will indicate human passage, but have not been productive in interdicting or apprehending those subjects. Some examples are outlined below:

- A high flying drone sees several suspect illegal immigrants; unless you actually see them cross the wire, how do you know they are not innocent farmers, etc.?
- Digital cameras are used for taking photos of forensic evidence; so what, the illegal has long since entered the country and disappeared?
- Modern smart phones can take pictures of a footprint and disseminate this to other tracking teams; to what aim, a person cannot be in two different places at the same time. Plus you are reliant on the mobile network actually giving you a signal.

Some modern technology such as GPS can help the visual tracker and also provide a distance between the tracking team and any illegal position. This will close the time lag of interception. In short, while technology is some help in establishing the overall picture, tracking teams perform better using their own skills. More information on various types of tracking technology can be found in Chapter Eight.

> **Author's Note:** Not all modern technology is on your side. In late 2013, a British Marine Sergeant serving in Afghanistan was jailed for life for his part in killing a wounded Taliban. The incident happened when the patrol came across a member of the Taliban who had been seriously wounded in an Apache helicopter strike. The Sergeant (name has been withheld to protect his family from reprisal) fired a round into the enemies chest, killing him instantly—or putting him out of his misery depending on your point of view. Unfortunately one of the other patrol members was wearing a 'headcam' which recoded the whole incident, including the voices. This footage, through a series of misfortunes, found itself in the hands of the British Military Police and the Sergeant was charged and found guilty of murder.

IEDs are one of the largest causes of casualties in Afghanistan, and spotting them is not easy.

AFGHANISTAN

I mention Afghanistan here because it is the current theatre of operations for both the American and British military. There has been some degree of tracking portrayed as the answer to countering Improvised Explosive Devices (IEDs), but sadly this is not the case.

The numerous amounts of IEDs in Afghanistan has caused many deaths and injured so many more. At one stage, it felt as if almost every patrol—both mobile and on foot—would get hit by an IED. They became so prevalent that the devices and the enemy got even cleverer. Seeing the impact of IEDs, they improved not just the devices but also the tactics, laying several IEDs in order to catch out more soldiers.

As a result of this, most infantry tactics changed, not to improve their operation capability, but to survive activating an IED. One of the major tactical changes was that most foot patrol walked in the footsteps of the man in front, literally forcing a single file approach to patrolling. If the soldier serving in Afghanistan learned anything about visual tracking it was to spot the signs of an IED.

INFILTRATION

Tracker teams are very often sent to a point where the enemy has been physically active in order to assess a good starting point. In many cases this can be a long way from their base and so a method of infiltration is used in order to get them to the start point as quickly as possible. Most militaries have the luxury of helicopters or fixed wing aircrafts from which men can be deployed by parachute. There are good and bad points for both. A helicopter will drop the tracker team exactly where they want to be, but the noise could alert the enemy. By comparison, an aircraft is less intrusive both in noise and view, but the parachutes can be seen if someone is looking up at the sky.

The infiltration method will depend very much on the terrain over which the tracking team will move and the amount of enemy presence and activity. Many teams are deployed by normal parachute entry and simply jumping out of an aircraft. At other times, a method known as HALO (High Altitude Low Opening) is used.

In a HALO drop, the parachutes do not open until approximately 2,500 feet above the ground. This requires the parachutist to free-fall for most of the way, a method of infiltration that is fast, silent, and accurate which tends to land the team in the same spot. The speed of descent in free-fall is fast, but may vary slightly with each individual and the position held. For example, in a normal *'delta'* position, the descent will be at a rate of 120 miles per hour, but in a *'tracking'* position this may well increase to 175 miles per hour.

Author's Note: During the heated debate when Guatemala threatened to invade Belize (a British protectorate), SAS patrols were inserted into the jungle to watch over the border area and establish any infiltration. I had a strong patrol of five men and we were given a region to watch in the south. Upon insertion, I immediately put *'Tom'* Bowler up front as lead scout with *'Merv'* Middleton to cover his back. While I was a fully qualified tracker, Tom had the advantage of being a real *'cool hand Luke.'* He had the ability to track and sense what was ahead of him and if Tom raised his hand you all stopped. On one occasion, Tom had thought he had heard voices. Investigation shows we were just a few meters from a jungle track and by now voices could be heard by all of us. I ordered an immediate ambush; followed by a quick message to base to standby for extraction should the enemy outnumber us. The *'enemy'* turned out to be local villagers returning home from Guatemala. I gave the signal and we emerged from the jungle line abreast catching the villagers completely by surprise; they were extremely distraught at seeing five heavily armed foreign soldiers. After a brief interrogation, we shook hands and let them go; hearts and minds play a big part in any conflict.

SUMMARY

Military tracking remains a skill, if a somewhat diminished one. It is a skill that relies on the individual understanding of what they are observing and interpreting that observation into worthwhile intelligence. Military tracking, when done professionally, is highly successful . . . and highly dangerous. The need for the tracking team to be out front and distant from support troops is vital to their success, thus they are required to have a high degree of self-discipline when operating in the field.

Those that have gone before us have proved that visual tracking can achieve real positive results. The Selous Scouts outwitted their enemy, as did the British in Borneo and the Americans in Vietnam. While technology might be improving at a swift rate, there is currently no technical devise that comes anywhere near the benefits of the human eyeball and an inquisitive brain.

In most military operations, it is the individuals and the team's nature that will get the job done. Looking after the man next to you, trusting in their skills and self-discipline all combine to make for a successful mission.

Chapter 6

TRACKING USING DOGS

Man has used dogs for military purposes for thousands of years. The Egyptians, Huns, and Romans all resorted to the use of guard and tracker dogs. Normally, dogs have been used for either tracking humans or, in more recent years, for assisting in the detection of improvised explosive devices. Border police also use dogs as a means of detecting drugs and other substances being carried by a person entering or leaving a country.

For the purpose of this book I will concentrate on tracking, patrol, and search dogs covering three basic functions carried out by dogs and their handlers. These include tracking, IED detection, and dog evasion.

Under no circumstances will a dog be used for tracking or any other type of military operation without its handler being present. The dog and handler are a highly trained team, and a dog cannot be handled by another person. The handler is an expert in his own field and will know the full potential and capabilities of his dog. The handler will also know under what conditions the dog is best used. However, the decision to use a dog as part of the tracking team is down to the team commander and not the dog handler.

War dogs have once again become part of the military establishment, as the threat of IEDs in Afghanistan has highlighted their importance in bomb detection. That said, dogs are subject to outside influences which have a direct bearing on their behavior. To obtain the maximum value from trained war dogs, it is critical to have an understanding of the conditions best suited for their employment. Being reliant solely on dogs and ignorant of their limitations can lead to a very dangerous state of affairs. It must also be remembered that a dog will tire easily and subsequently its advantage to the team will be short lived.

The handler's harmony with its dog is also of vital importance as they must understand the dog's limitations and read what the dog is telling them. It is possible for the handler to pull the dog off the scent believing that the animal is wrong where, in reality, the dog is correct.

In recent years, war dogs have once more proved how invaluable they can be. When properly used, they can provide a massive advantage over the enemy. This is particularly true in war zones where dogs are not tolerated by the local populous, such as the Middle East. In countries where dogs are household pets, they can pose a massive problem because they are the ultimate deterrent against burglars and unwanted people visiting your home.

Most people think that dogs are used mainly for mountain rescue or by the military—they would be wrong. In addition to tracker dogs, dogs that search for IEDs and patrol military bases for security, man's best friend is a valuable asset to society and gets little credit for it. Dogs can be trained to carry out a whole variety of roles:

- **Customs and Border security:** Literally hundreds of thousands of dogs are used at airports and shipping docks to search for contraband, explosives, and narcotics. Their success rate is much higher than that of humans.
- **Urban disaster:** The most difficult SAR specialty, urban disaster dogs search for human survivors in collapsed buildings. They must navigate dangerous, unstable terrain, and can crawl where humans cannot gain access.
- **Dead body:** Dogs specifically search for the scent of human remains, detecting the smell of human decomposition gasses. Cadaver dogs can find something as small as a human tooth or a single drop of blood.
- **Avalanche:** Dogs search for the scent of human beings buried beneath up to 15 feet of snow.
- **Wilderness:** Dogs search for human scent in a wilderness setting.
- **Water:** Dogs search for drowning victims by boat. When a body is under water, skin particles, and gases rise to the surface, so dogs can smell a body even when it's completely immersed.

A Deutsch Langhaar is bred for hunting and tracking which works particularly well in water.

SCENT TRACKING

Scent tracking is probably the most widely accepted method of tracking to pursue human beings. Scent dogs that have been used primarily in the law enforcement and search and rescue community are now being used heavily in the military's CIED initiative. However, while tracking dogs have proven to be an indispensible asset, there are some disadvantages associated with using dogs for visual tracking. The most difficult job a dog can perform is to follow a scent trail. The level of effort is so intense that most dogs cannot work longer than twenty to thirty minutes at a time, followed by a ten to twenty minute rest period. At best, dogs can perform this cycle no more than five or six times in a twenty-four-hour period before reaching complete exhaustion. The efficiency of the search also decreases as the dog tires. If the subject keeps moving and stays out of the detection range of the dog and handler, the subject can outlast scent tracking dogs.

The trackers also credit the dogs for repeatedly saving their lives—alerting them to enemy soldiers and to the booby traps that the North Vietnamese were known for leaving behind. The British developed Labrador retrievers as jungle trackers in part because the easygoing dogs could adapt to different handlers as soldiers rotated out of teams at the end of their combat tours, Merritt says.

"There is something about a Lab that makes it a great tracker," says Taitano, who was the least likely dog handler. He grew up in Guam, where dogs are considered a nuisance, not pets or working animals. "When we see a dog in Guam, we kill it," he says.

But a black Lab named Moose captured Taitano's heart. "I had a best friend," he says. "It is a privilege to work with a dog who gives you unconditional love." He tried to buy Moose at the end of his tour so he could take his tracking companion home, but the Army would not allow the dogs to leave Vietnam because of the risk of spreading diseases they acquired in South East Asia. That still bothers the trackers.

"The terrible thing is, the dogs were left behind—and the Vietnamese eat dog," Peterson says. "That was always gut wrenching."

SENSORY CHARACTERISTICS

The dog relies very little on sight during its day-to-day activities. Its attention is, however, drawn by sound and movement and if its interest is aroused, will follow up with hearing and smell. There is no evidence that dogs' night vision is any better than man's although its low position may help, as more of what it looks at is better defined.

A dog's hearing is twice that of a human and a dog can be attracted by a noise not heard by the handler. This is an important facet that any tracking dog handler should note, as the dog often becomes aware of the quarry's presence before the handler. The distance at which sound is received is very much governed by the weather, in particular wind and rain.

The dog's main facility over that of humans is smell. It is estimated to be some seven to nine hundred times greater. It can track microscopic traces of substance or their vapor that lingers in the air, on the ground, or that has become in contact with other objects. The dog can also detect minute disturbances on the ground, which may cause an alteration in the scent picture. Humans have no equivalent to a dog's scent picture, as the animal actually thinks in smells. It pictures one of an object itself either animate or inanimate associated with the smell from human perspiration, to toiletries, to the smoke from a cigarette or the type of polish on a soldier's boot. These are mixed with the smells of a boot crushing leaves or dust thrown up from someone walking over a dry patch of earth.

THE SCENT PICTURE

The scent picture is analyzed in two ways. These are known as the air scent and the ground scent. Air scent comprises mainly of body scent, clothing, deodorants, toiletries, and even chemical aid that is used in washing clothes. Most of all a dog can detect the smell given off by an individual human within a group of smells. The total amount of body scent given off by a human will depend on constitution, activity, and mental state. A quarry which moves or stands still leaves their scent suspended in the air for a short while before falling to the ground. Prevailing weather, especially the wind, has a great effect on this scent.

A human nose has 5 million scent detecting cells, while the Bloodhound's nose has some 300 million. The dogs nose has a pattern of ridges and dimples that, in combination with the outline of its nostril openings, make up a *nose print* as unique as any human fingerprint.

From the dog's view, any ground scent deposited will consist of two pictures; one is the body scent of the individual and the other is the disturbance made in the environment as each foot hits the ground. This causes crushed vegetation, dead insects, and the breaking of ground surface, which releases a gas vapor. Ground scent can last up to forty-eight hours and even longer in ideal conditions.

Factors that will affect the scent picture are moist ground conditions, vegetation, humidity, forest areas, and light rain mist or fog all act to make favorable scent conditions.

FACTORS THAT AFFECT SCENT

All scent is subject to decay and can be affected by a range of natural factors; the main three being the terrain, weather, and time. Scent is adversely depleted in arid areas of sand, stone, paved roads, and city streets. In high winds and heavy rain, the scent is rapidly diluted; strong sunlight will help increase evaporation of scent. The older the scent picture, the harder it will be to follow.

However, light warm winds can draw the scent smell and improve the dog's scent picture. Snow and ice can also hold smell, releasing it only when it thaws. Many people believe that substances such as pepper will clog up the dog's nose; this is untrue and the pepper only serves to enhance the dog's scent picture. Some substances, such as potassium permanganate, will severely affect any dog's ability, but getting the dog to inhale such a substance would be difficult.

Many people have tried covering themselves in animal excrement in order to avoid being detected by a dog—the dog will simply smell a human covered in animal excrement. A dog will be distracted if he smells the scent of a bitch in heat, as it will prefer to follow this scent than that of a human; however this is rare and the chances of it helping a quarry being tracked is highly improbable. There are repellents that smell the same as a bitch in heat and these will slow the dog down if used.

Decoy scents may be used which are stronger than the human smell the dog will identify as the prime scent to follow. Dragging a potential food source the dog likes such as a rabbit or dead fox may help break the scent trail.

DOG CAPABILITY

The capability of a tracker dog to complete a successful track will be dependent on many things, no more so than the dog's own ability and the success in dog tracking training. Each dog will respond differently to its training, so proving vastly more superior than others. The strength and stamina of the dog is also extremely important, as they can only be effectively worked for short periods of time. Factors governing the tracker dog's capabilities involve:

- The quality and distribution of the scent.
- The age of the scent.
- Climate conditions.
- Other distractive scents.
- Heat—that causes rapid evaporation of the scent.
- Extremely rugged terrain.
- An Unverified Start Point at an incident site will cause a dog to follow the wrong route or scent.
- Low humidity causes the scent to disperse more rapidly.
- Dry ground does not preserve the scent.
- Wind may disperse scent that will cause the dog to track downwind.
- Heavy rain will wash the scent away.
- Distracting scents will take the dog's attention away from the trail. (Some of these scents are blood, meat, manure, farmland, and populated areas.)

- The scent becomes covered by elements of nature and cause the scent picture to be partially or completely covered. Examples are sand that can blow over the tracks and help to disguise the track; snow and ice that can form over the track and make it nearly impossible to follow; and water which is one of the most difficult conditions for a tracker dog team.

With a good scent, a well-trained tracker dog can expect to locate its quarry. Nevertheless, the factors above will limit its ability. If the scent is old, it will have dissipated; heavy rain will also dampen the scent and the dog will become confused if the tracking scent becomes mingled with other odors.

Normally under good conditions, a good scent picture that is less than one day old, will allow a tracking team to cover up to 5,000/8,000 meters depending on terrain. Providing the tracking team stay on the trail and the gap between the quarry and the tracking team is not extended too far, there is no reason why they should not catch the quarry.

One vital part of any tracking dog training is the capability to alert the handler to the presence of an ambush or booby trap. This ability to detect such things as immediate human presence, unnatural disturbances, and trip wires is one of the best defenses a tracking team can have. Training the dog to be an early warning system is indispensable.

One of the greatest causes for concern is the inability of the dog handler not to have full confidence in the dog's abilities. In circumstances where the handler will force the dog off the scent because he believes the dog is wrong will only serve to destabilize the animal and its skills. It is my personal experience when working with tracking dogs, no matter where they lead the handler, nine times out of ten the animal is right.

Before any dog can be chosen for specialist training, it must undergo the normal basic obedience courses. The handler must become solid friends with his dog, spending as much time as possible together. The handler must be aware of all the requisites in preparing a dog for tracking training and be able to look after the dog once on operations.

It is always better to get a dog when it is fairly young; just like people, the young adapt quicker. It is difficult to say who chooses who, but normally the military dog handler will have a basic love of all dogs. Dog's sense love and trust very quickly and in many cases, the handler and dog become an inseparable team.

Dogs like a cuddle; they love to be touched and hear words of encouragement; and will never get enough of it. A good dog will be obedient; for example when walking it will stay at your side and not try to pull ahead. Dogs like food and treats, but they should not beg or come around when the handler is eating. Some dogs can be extremely jealous of other people or animals, so the best dog is one that is comfortable with other members of your tracking team. That said, it is imperative that the handler be prime and that the dog does the handler's bidding not that of the other team members.

A war dog with handler.

Photo courtesy of Petty Officer 2nd Class Todd Frantom

A GOOD HANDLER

The bond between handler and dog can be increased with time. While the dog is undergoing its tracking training, the handler should also learn how to look after the dog during operations. This will require having knowledge about dog aliments, its nutritional requirements, and its resting habits. Most of all, the handler must learn to recognize the dog's actions. For example, when a dog quickly raises its head and looks in a certain direction, it has most probably heard something the team has not. Interpreting the head movements and throat sounds a dog makes provides the handler with important messages.

Successful care and training of dogs depends, to a great extent, on the personal characteristics of the handler. They must show patience and persistence in their training of the animal and not expect the dog to learn as quickly as humans. The handler also needs to be as physically fit as his dog, making sure they can keep up and operate together. The life of a tracking dog requires a lot of endurance training and long walks over rough terrain. Most dogs can swim well and the handler should encourage the dog by swimming with it.

Finally, remember that the dog cannot talk and so it is sometimes difficult to understand what the dog is trying to tell its handler. This telemetry between the handler and dog can be improved with the handler paying attention to the dog's welfare at all times. The dog must be kept clean and well brushed; its harness and leads must fit well and not hurt the animal in any way. If the dog requires basic first aid, then where possible, this should be done by the handler. Feeding the dog both in the camp base area and during operations will also strengthen the bond between handler and dog.

Dog Training

Photo courtesy of DoD Master Sgt. Cohen Young

TRACKING TRAINING

The training of a tracker dog is not unlike that of a human as mentioned in Chapters Three and Four. The terrain over which the dog and handler move can vary slightly and those laying the track may move through the bush as opposed to down a track. Again, as with humans, there is little point in taking a new dog out for training if the weather is extremely windy or raining heavily as this will result in the animal getting confused and learning little. Advanced training in severe weather conditions can take place once the dog and handler have gained confidence in each other.

Tracker dog training should take part over similar terrain to that which the tracker team will be operating if possible. While this is not always practical, at the very least the dog should be acclimatized to the area over which they will be trained. It requires several days for the dog to adjust the ambient air temperature, soil type, vegetation, and regular odors that emit in everyday life. This will help lay a clean canvas in the dog's mind, making it more responsive to any fresh scent.

Normal forest and farmland are good areas to start, while swamps and wetlands should be avoided in the early stages. As the animal progresses, it can be moved into new areas closely resembling the operation terrain. This is vital if the dog is expected to track over harsh desert or arctic areas. Some ideal tracking training conditions will include:

- Thick vegetation with good ground cover will help limit the dissemination of the scent.
- Cool, cloudy weather which will limit the amount of scent evaporation.
- Training early morning and late evening is also recommended as this again reduces scent loss.

During this training the handler must also learn what the dog is doing. For example, the handler may well see a footprint which is conclusive sign, yet the dog may be off several meters to one side due to the wind pushing the scent. Additionally, the dog may sometimes move ahead faster and the handler must interpret this. A dirty quarry leaves more scent, as does a person moving quickly.

As with human visual tracking, the dog needs a start point: in the early stages of training it is best to provide the dog with some clothing belonging to the quarry so that they have a definitive scent to track.

Scent picture is a combination of not just the individual smell, but also of additional odors coming off the quarry. For example, an enemy soldier may be a heavy smoker of a particular cigarette—this will leave a reinforcing scent for the dog to follow. Additionally, as the quarry moves through the vegetation, a vast amount of scent is thrown into the air from crushed insects, broken branches, and trodden grass. The combination will provide the dog with a good picture of the trail the quarry has left behind.

All military dogs are carefully selected, for their temperament and breed ability as some dogs are better at certain tasks than others. The types of war dogs that are in common use are:

- Patrol dog
- Tracking dog
- Mine detection dog

THE PATROL DOG

The patrol dog can be used for a wide variety of tasks, from reconnaissance to fighting patrols. A normal patrol dog works by 'air scent' and hearing. Upon detection of either, it will provide a silent warning to its handler. They can be on a lead or free, in either case they will be controlled by the handler. When kept at 'heel' the dog is being held back—when released or put on a search rope the animal recognizes it is to go on alert—words of command will be used such as 'seek.'

The dog will provide sign of an intruder by raising its head and picking up its ears at the same time, and generally dependant on the strength of the scent, the dog's body may become tense. Other signs such as tail wagging or the pull ahead to investigate may be present. While it is not a tracker dog, it can provide ample warning of a possible ambush or at camp infiltration. Either way, it is better at detecting a human at extended distances. Most patrol dogs can work by day and night though their effectiveness is governed by several factors.

- The wind direction and velocity.
- Concentration of the scent.
- Distractions within the vicinity, such as noise, light.
- Blocking effects of humidity and thick vegetation.
- The dog's natural ability.
- Most importantly, the handler's ability to recognize the dog's warnings.

LIMITATIONS

- Other dogs in close proximity will distract the animal.
- Dogs do not recognize the difference between one human and another. However, they can be trained to be discretionary with certain scents given off by people.
- The dog will become confused when large numbers of people are in a small area or when contact is made and there is lots of sudden gunfire.
- In addition to battle sounds and smells, most dogs will become distracted when both friendly and enemy forces start running around in different directions.
- Dogs will obey their handler even if he is wrong in his assumption.

Labrador retriever mix who is deployed with troops in 64th Brigade Support Battalion, 3rd Brigade Combat Team, 4th Infantry Division, Multinational Division Baghdad.

Photo courtesy of DoD Pft. April Campbell

THE TRACKER DOG

Tracker dogs may be used to help track an enemy. Tracker dogs are trained and used by their handlers. A dog tracks human scent and the scent of disturbed vegetation caused by man's passing.

Tracker dogs should be used with tracker teams. The team can track visually and the dog and handler can follow. If the team loses the signs, then the dog can take over. A dog can track faster than a man tracks and can do so at night. A tracker dog is trained not to bark and give away the team; it is also trained to avoid baits, cover odors, and deodorants used to throw it off the track. Dogs used for military purposes generally conform to certain requirements, irrespective of its breed. Below is a guide.

Physical (Height to Shoulder)	22 to 26 in.
Weight	45 to 100 lbs.
Speed	25 to 30 mph for a short distance

Their temperament is also vitally important. The dogs should be intelligent, courageous, faithful, adaptable, and energetic. A dog must heed the commands of its handler, yet when instructed to do so, become fairly aggressive. A well trained dog is more than capable of taking down the average man or woman. While the type of dogs used will depend partly on the usage, guard dogs which protect installations and military compounds may differ from those used for tracking or IED detection. Some of the best breeds used are:

- Alsatian
- Doberman
- Rottweiler
- Mastiff
- Labrador

The tracker dog is employed to find and follow the scent of a man's progress on foot. They work mainly on ground scent, unlike the guard dog, which would primarily work on air scent and noise. Normally the dog will follow the freshest scent and a lot depends on teamwork between the dog and the handler to achieve the best results. The dog is trained to follow a distinct track; it is up to the handler to make sure that it is the correct one. In the case where there is dispute between the dog and the handler, the dog is normally right, as the handler can be deceived by the quarry using anti-tracking techniques.

Tracker dogs are trained to follow human scent and can identify any individual by their unique smell. They generally work on a food reward basis and are given food after they have returned from the day's tracking. The animal operates best when the air and ground temperatures are roughly equal. It also helps the dog if the there is a certain amount of moisture in the air which will help slow evaporation of the scent. Ideally when tracking through thick vegetation the scent will remain strong and will also be helped by the amounts of damaged vegetation left behind by the quarry. The more scent the dog can detect the better the chances of a quick, clean follow. This depends on a number of factors:

- A person running gives off more body odor than moving calmly.
- A person who is unclean or bathed in a distinctive scent such as strong soap or after shave.
- Damp ground and vegetation will help preserve the scent.
- The smell of fresh blood is very easy for a dog to follow.

However, there are also many factors that adversely affect tracking these include:

- Hot dry sunny exposed areas.
- Concrete and tarmac areas such as roads and streets.
- Strong winds which can blow the scent off the line of travel.
- Heavy rain will also erode the scent as will running water, rivers, etc.
- In areas where there is heavy animal scent, large wild herds or domestic animals such as cows and sheep.
- The downdraft of a helicopter approaching any incident scene can abruptly destroy any scent.

Where possible, the dog should be given the best and freshest scent possible; any discarded clothing or equipment that can identify the quarry is invaluable. It is best if the dog can be brought to the start point of the track or where an incident or engagement took place. The quicker this can be done the better the end results, as the dog will have a much clearer scent picture to work with. If a tracking dog is to be deployed then all efforts to preserve the scent and avoid contaminate should be taken. Additionally, when tracking an enemy, it is best to keep the presence of a tracking dog a secret as far a possible to avoid the enemy taking evasive actions.

During any daytime track, the handler can see what his dog is doing and read the signs provided by the animal. While tracking dogs can be used at night with a great degree of success, there are several drawbacks to this. If the tracking terrain is fairly open bush or flat lands then the tracking team will be able to keep up. However the dog, even on a lead, can move ahead quickly when following a good scent and over rough terrain this can cause the team to become fragmentized. Additionally, when moving in the dark, it is difficult for the handler to see or anticipate all the signs the dog might be sending out. One of the greatest dangers of night tracking is walking into an enemy ambush, especially as the early hours of dawn approach.

In a military situation where the enemy has attacked during the hours of darkness and a follow up has been authorized, it is best to let the tracking dog only work for a few hours. This should establish a direction of travel and leave sufficient scent for the dog to continue early the next day during daylight hours. Establishing the direction of travel will also allow the unit commander to put blocking troops in place ahead of the fleeing enemy.

TRACKER DOG TEAMS

While it differs from army to army, military tracker dog teams usually consist of two or three dogs at best. Along with the handlers, there will also be good communications to allow any follow up team to support the tracking unit. The dog team will normally have the support of infantry in order to protect them from ambush. Tracks over open ground may also deploy a sniper with the dog team; if spotted, the quarry can be shot at long range when seen. Long tracks over several days will normally use two dog teams rotating as required.

All dogs react differently and it is up to the handler to understand and appreciate their own dog's traits. Many handlers and dogs become friends for life and there are cases where when one has died, the other has reacted as if they have lost their best friend.

L/Cpl Tasker and his faithful friend Theo both died on the same day.

Author's Note: On March 1, 2011, twenty-six-year-old Lance Corporal Liam Tasker, a dog handler with the Royal Army Veterinary Corps, was killed in a fire fight with insurgents in Helmand Province, Afghanistan. A few hours later his faithful dog, Theo, died of a heart attack. Theo was a Springer Spaniel trained as a sniffer dog to detect roadside IEDs. So strong was the bond between man and dog that the death of Liam brought on a severe heart attack. Both soldier and dog were repatriated to the UK and both received recognition of their bravery.

IED Dog Detection

Photo courtesy of Dod Sgt. Alfred V. Lopez

IED DETECTION DOG

In the last fifty years, there has been a massive increase in the use of IEDs. From the troubles in Vietnam and Northern Ireland to Iraq and Afghanistan, IEDs have rapidly became one of the most effective anti-personnel weapons available to enemy forces and insurgents, resulting in thousands of casualties to date. While not dissimilar to the tracking dog, this animal is trained to detect booby traps, mines, hides, or ordnance caches. Once a mine has been discovered, the dog will normally sit down within a few feet of the object thus informing the handler of a possible problem.

In addition to the human cost, the medical commitment, and millions of dollars in expensive equipment, it has become one of the main features of war. To combat the threat, millions of dollars were spent on IED-detection equipment and research, yet while this expensive equipment made some difference it was no were near as reliable as the humble dog. Time has shown that no robotic machine can compare to the performance capability of an IED trained dog.

- The success rate of a dog is extremely high, bolstering the confidence of troops on the ground.
- Where equipment is usually component-specific in its detection, a dog shows no limit to the amount of explosive materials it can detect.

- The dog can work with the men on the ground as part of the team and does not have to be transported to and from every incident.
- The cost of one IED trained dog is minimal compared to the cost of developing and purchasing IED robotic equipment.
- Moreover, they make for a safer patrol; restore confidence; and limit the amount of soldiers returning as limbless casualties.

DOG EVASION

For any quarry being followed, the threat of dogs comes from two directions. First there is the inadvertent compromise of one's presence by dogs from any source, such as normal domestic animals, especially in rural areas or farming country. Secondly, and by far the biggest threat, is detection, pursuit, and discovery by professional tracker dogs and their handlers.

Evading a dog which is tracking you is not easy. If you hear a dog bark then you should stand still and assess which direction it is coming from. Handlers will sometimes make the dog bark in order to hurry their quarry into making irrational decisions or to scare them into movement. If a dog has spotted a moving man, the animal may lose interest if the man 'freezes.' In the case where the tracking group is in immediate pursuit and close by, there is not much one can do but defend one's self. In a delayed pursuit—even if the delay is very short—there are several counter measures that can be taken. The main factor here being the distance between the tracker group and the quarry. The idea being to maintain or increase the distance between dog and quarry; I mention dog and quarry as opposed to tracker and quarry, as in some cases it is possible to fool the tracker, but not the dog following the quarry. The main points are:

- Run steadily.
- Climb up or jump down vertical features.
- Swim rivers.
- If the quarry is in a group, split up.
- Run down wind.
- Do things to confuse the handler.

- Mix with domestic animals if nearby. (Strong smelling animals such as cattle will aid the quarry, throwing the dog of their scent.)
- Change your footwear in areas where there are multi-tracks being made by others, this will confuse the handler, but not the dog.

Author's Note: If you are being tracked by a dog and handler, the quarry should cross a major obstacle such as a river, walk some two hundred meters downstream and cross back over the river. This can be enhanced by climbing up a steep incline when it is not necessary or doing something totally out of character. If this pattern is repeated several times, the handler will think the dog has lost you and call the dog off, whereas all you have done is confuse the handler.

When attacked by a dog, don't run, face the animal and offer it something to grab onto.

ATTACKED BY A DOG

While very few tracker dogs will actually attack a quarry, there are some that sense the needs of their handler and as such will aim to bring the quarry down. An attacking dog will attempt to paw down any barrier placed in front of him, so using a strong stick to bar his path could help. Normally the dog will wish to take a bite and *'lock-on'* to the prisoner. If this is the case, wrap a coat or sack around your arm and offer this to the dog. Once the dog has taken a grip, stab it in the chest or beat it on the head with a rock or stick. Make sure that whatever you do to the dog, it is permanent; otherwise it will just become even more annoyed. If the handler is not present and you have no other weapon, try charging directly at the dog with your arms outstretched and screaming. Given the size of a human in regards to that of a dog and the sudden unexpected nature of the attack, the dog may break (do not rely on it), as a dog's confidence and security can be weakened very quickly.

Do not try to use any chemical substances—such as pepper—to put the dog off the scent, as this only increases the scent picture. If you are fairly fit, remember the dog is only as fast as the handler—you should always try to outrun them. If cornered by both dog and handler, it is best to give up unless you are armed. Killing the handler's dog could only bring distress upon you once recaptured. If you are armed, then choose a location that is best suited to ambush; in such a case kill any human first as they too may be armed—deal with the dog as a last resort.

When a dog is charging at you, try to break its momentum on which it will rely to knock you to the ground. This can be achieved by standing exposed next to some object like a tree until the dog is a few feet away, in the last second move rapidly behind the tree. The dog is forced to slow in order to turn—take advantage of this.

Other things to take into consideration is what it takes to support that canine, such as carrying food and water for the dog on long operations. Canines are great for alerting a patrol as to human presence. However, a dog can also give away that friendly patrol location as well.

SUMMARY

There is no doubt that dogs play a major part in tracking. Given their extraordinary sense of smell and acute hearing they are able to build a picture of the quarry humans could never hope to achieve. The diversity of skills provided by dogs and their ability to adapt to the demands of their human masters is almost limitless.

The bond between man and dog is an extraordinary emotion, and if we are fair they treat us far better than we treat them. Dogs will willingly risk their own lives to save that of a human, and are faithful to the point absurdity, displaying undying devotion for their master.

Apart from being a valuable search asset, dogs also offer the best protect anyone could wish for. Even a normal domestic house dog—no matter the size—will protect your home far greater that most electronic devices. The dog's keen sense of smell and hearing will set him to barking and thus create an alert that rarely goes unnoticed.

Author's Note: One of the major problems the British SAS had in Northern Ireland during the troubles between the 1970/80s was the inability to get close to suspected targets' homes. In the countryside, many homes employed dogs which would detect the presence of surveillance parties trying to get in close for reconnaissance. No matter what equipment and chemicals were available to the Special Forces, dogs remained one of the hardest problems to overcome.

Chapter 7

ANIMAL TRACKING

Animal tracking is done for many reasons. While we think primarily of hunters tracking animals, more often than not it is a dedicated group or organization checking up on the health of a certain species. Additionally, many people simply go out tracking in the wild for the pleasure of tracking and locating an animal to simply photograph.

The main difference between human and animal tracking is the diversity of animal species. Where humans may be a different size or weight, they still walk on their feet—with or without footwear. By comparison, there are thousands of animals all with different shaped feet, size, and patterns. Some live on land, others live in water, while some have their habitat in the trees.

There are similarities between human and animal tracking, as both leave sign behind. But animals do not try to leave a deception behind and, secondly, animals have a much better sense of human approach. Additionally the sign left behind by animals can be far greater than those left by humans as they are left naturally, but sometimes they are obscure to the unprofessional eye. Animal tracking involves locating and interpreting many facets such as:

- Animal behavior
- Tracks and prints
- Track pattern and gaits
- Territorial feeding sign
- Bedding sign
- Feces and droppings
- Territorial marking scents
- Mating scents
- Environmental disturbance
- Animal tracks are confusing

I would be amiss to understate the difference between human and animal tracking. Unlike humans, which have a fairly standard footfall pattern, in the animal kingdom there are thousands of different patterns. Some do look similar, while others can be completely baffling. Humans can also become confused over what they are actually following unless they have a good insight as to the animals in the local vicinity.

One animal can make a track and yet it can look like a completely different animal footprint depending on the surface the track is imprinted upon. For instance: sand, mud, and snow will alter the image of an animal's footprint. You are also rarely going to get a perfect print with the elements against you, i.e, snow thaws and rain will wash away mud, both resulting in a distorted image. Even if you have overcome these obstacles and are con-

vinced you are trailing one type of animal, it could still turn out to be another. Many young animals can leave a print which is not too dissimilar to that of that of a smaller species. Likewise, tracks made by forefeet and hind feet can once again trick you into thinking you are following the tracks of a different animal. With all this against you, you need to have a clear idea of what to look for in the first place. Your conclusion should not be based purely on the print, but also on your surroundings. You should be looking for what type of substance the track has been made in, what time of day or night it is, what are the weather conditions (could they have affected the print), and most important of all, what are your surroundings.

For a hunter, understanding animal tracks is vital to locating his quarry.

As an example of a similar track, I am going to use a rabbit and a squirrel. The rabbit will push off with its hind feet and land on its forefeet, which touch the ground one after the other. The hind feet then touch the ground landing in front of the forefeet. This leaves a print of the larger hind feet, followed by the print of the smaller front feet. The squirrel has a similar style of movement with the hind feet landing in front of the forefeet, leaving the same type of print. Being so similar, it would be very difficult to decipher which footprints had been left by which animal if it was not for one simple solution: a squirrel's trail starts and ends at a tree.

Animal behavior is very diverse. As seen in the image above, this male and female lion are in rapture with their eyes closed, a behavior trait similar to humans.

A series of tracks made by an animal will give you a trail. This gives you an idea of the speed an animal was going. The greater the gap is between the group of tracks, the faster the movement of the animal.

A walking animal moves its right forefoot first, followed by the left hind foot. Then the left forefoot is moved, followed by the right hind foot and so on. A trail made by a walking badger will show that the hind foot has landed on the track of the forefoot. This is called being in register and is what happens when a badger (or any other walking animal) has been walking or trotting; it moves its legs in a definite order. If an animal has been galloping or jumping, the tracks will not be registered in a normal pattern.

ANIMAL BEHAVIOR

One of the most important details to register in your mind when tracking an animal is its normal behavior. All animals have to drink, eat, bed down, reproduce, and in some cases hunt other animals. Some adult animals will claim a given territory and defend it with their lives, as within this territory they have water, food, and a place to mate and bring up their young.

Life for most animals is extremely difficult and a constant worry, but in their natural habitat and unmolested, they will find a certain amount of comfort and so settle into a daily routine very much as humans do. In such a case, they will routinely take the same path to water, feed on the same vegetation, and frequently bed down in the same spot.

We often think of animals as being fairly dumb, but they possess much of what man has lost: the will and skills to survive. Animals are extremely aware of any interruptions or disturbances within their territory. And when disturbed, most will have already chosen an escape route or a bolt hole in which to safely hide.

Studying the behavior of a particular species will help in locating and tracking an animal. Knowing and understanding the environmental requirements of each species will save a lot of time when looking for a start point.

TRACKS AND PRINTS

As with human tracking, animal tracks and prints are fundamental to any professional tracker, as from these they can identify the species, size, age, weight, and sex. There is a fairly comprehensive list of many animals further on in this chapter.

Both front and rear tracks are easy to define as many have larger forefeet, especially the fast runners. Additionally, most females of the species are smaller than the males, yet they can move almost as quickly.

The first thing for any tracker to do is locate a clear print (study the pad and toe configuration). As many prints can look similar at first glance, a professional tracker will be able to identify is the species finding its food by

digging; chasing other animals, or climbing trees. The number of toes is also a valuable indicator as most carnivores are designed to run fast and have four toes. The slower moving animals—such as bears—have five toes on all four feet, while other smaller animals—such as squirrels—have four toes on the front and five toes on their rear feet.

While the list of animal tracks and prints is long, it is always advisable in the beginning to carry a small book which describes the various types of sign you might come across. Understanding what you are looking at when you find an animal print is a great place to start.

You need a lot of experience to fully understand even a small amount of animal tracks and prints, but this dog (left) and Moose (right) are fairly easy to identify.

TRACK PATTERN AND GAITS

It is not just the animal print that is important; finding several prints in the same area will also help you identify the track pattern. Some animals walk, others trot, and some hop. At times they also run . . . especially when taking flight.

Animals always look where they are going and will place their forefeet safely down where they can see them. As a matter of course, most animals bring their rear feet up to almost the same spot because they know it is safe to do so, while some species will swing their rear legs up and past the front legs in a swinging motion.

The gait made by each animal can be measured both directionally and from left to right. This will help understand the size of the animal and how wide it is, thus giving some indication as to age.

Be aware that at times both front and rear prints can be in the same spot making it a little confusing for the tracker. Moreover, the terrain over which the animal is moving will also impact on the prints left behind; whilst mud and light snow will reveal fairly good animal prints, grasslands and forest floors make it difficult to determine an exact print type.

FEEDING SIGN

Finding animals is easier if you know and understand how and what they eat, that way you can identify the food source and then find the animal. There is a saying, "we are what we eat," and the same is true for wildlife. The best time to locate a feeding ground is when the fruits and nuts are out on the trees and bushes. One of the simplest ways to start looking is to identify the food groups we already know, and then associate these with which animals eat them. For example, we all know what a hazelnut looks like and could recognize it in its natural environment. So what animals eat hazelnuts?

- **Squirrels** tend to split the nuts neatly in half.
- **Common dormice** leave a smooth, round hole in the side of the nut, with tooth marks running around the inside of the hole.
- **Wood mice** leave tooth marks on the surface of the nut and across the edge of the hole. The hole may be either circular or ragged in shape.
- **Bank voles** create a round hole with tooth marks across the edge, but not on the surface of the nut.
- **Great spotted woodpeckers** break nuts into pieces or leave large, irregular holes.

- **Woodpeckers and nuthatches** also jam hazelnuts into crevices in tree bark, so they can hammer them more easily.

If you adapt the same strategy to larger animals, you would find that deer will skin with their cheek and teeth the skin of a conker (horse chestnut), leaving the skin on the ground. Many carnivores eat fruit in the autumn which can be seen quite clearly in their droppings. Sometimes slugs and snails are drawn to the droppings as they are attracted to the digested fruit. Wolves and foxes will leave wild blackberry pips in their dropping, detectable by the strong odor of the wolf droppings.

BEDDING SIGN

As with humans, animals will tend to live their lives out in the same area. They will have the same sleeping area, travel routes, and feeing areas. Added to which—just like humans—they will take the easiest and most logical route to and from their feeding areas and bedding.

Bedding areas will differ slightly, but most animals such as deer and wolves will simply lay down on a nice soft spot; in many cases this will be under an overhanging branch. While bears and cats will also sleep out in the open, as the weather gets colder they tend to find deeper shelter in caves, fallen tree stumps, or simply burrow deep into the undergrowth.

Trying to spot the sleeping area of a wild animal is not as easy as one would think. First you must research the animal species you wish to locate and learn its individual habits. There are a few exceptions such as the Roe deer, as this small animal likes to clear away any leaves and vegetation and lay next to the earth. This makes its bed area easy to spot if you know what you should be looking for.

SCAT, FECES, AND DROPPINGS

As far as visual tracking goes, animals have one major sign over humans: they leave behind their droppings. These are referred to as *'scats'* in many books and manuals on animal tracking, but do not be confused; it simply means that anything the animal has eaten and which in time passes through its digestive system. In many cases the word scat is retained for certain animals whose droppings

are naturally encased in twisted fur; animals such as wolves and cougars leave scats behind. The reason for this is simply to protect the animal because it eats bones which splinter and could prove dangerous.

Finding any dropping is a great benefit to the tracker as it provides conclusive sign of the animal in the area. Further examination of the dropping will reveal what the animal has been eating. Animal droppings decay with age and the experienced tracker will be able to define how old the droppings are by breaking one open. The longer they are exposed to weather and sunlight, the more decomposed they become.

An animal does not just do its "business" in any old place when the urge takes them; they use their droppings and urine to indicate territorial limits. The droppings serve as a sign post to other animals both by smell, size, and the gender of the animal. While humans will simply see a pile of droppings or a place where the animal has recently urinated, other animals will absorb a lot more information via their olfactory senses.

Apart from the age of the droppings the type of information a human will normally pick up is what the animal has been eating. While some droppings from different animals can look the same, the content may vary. For example, in carnivores you are more likely to find bits of bone from other digested animals, while in herbivores you may well find the seeds of berries they have eaten. This will help you identify the animal group:

- Carnivores: flesh eaters
- Omnivores: eat both flesh and plant life
- Herbivores: eat only plant life.

Scats and droppings will tell you about what and where an animal is feeding.

For example, if you come across what looks like a cow-pat in the forests of Canada, it is most likely an elk or a moose. On the other hand, small fur-wrapped scats which contain minute bone fragments which may indicate a small species of cat that has been surviving on a diet of rodents.

Many animal trackers will place more emphasis on the animal droppings than the actual print the animal has left behind. The study of animal droppings cannot be overstated because they not only provide conclusive prove of the animal in the vicinity, but help indicate where the animal has been feeding. And to the professional eye, may also indicate the age, size, and health of the animal.

Some animals are carnivore while others are herbivores.

TERRITORIAL MARKING SCENTS

Animals mark their territory for a number of reasons, but mainly to reserve an area of ground over which they can claim dominance. This area is generally reserved to protect the females and any young; to provide feeding and access to water; and to bring certain calmness to the animal so it can bed down without being under constant threat.

While we have seen that animals use both their droppings and urine to mark their territory, they also leave a wide variety of other sign; from claw marking trees to digging up the earth. Tree markings are sometimes mentioned to as 'scratch poles,' which to the trained eye are fairly common in the woodland areas festooned with certain animals such as bears and deer. Whatever each animal uses, be it scent, ground, or tree marking, it is a clear indication to other animals that this territory is occupied.

MATING SCENTS

All animals mate and in order to do so most have developed a 'mating system' designed to attract the opposite sex. While not a written rule, most males will pursue females, the latter generally keeping a low profile and concentrating on feeding and getting ready for the pregnancy. Additionally, almost all mating is seasonal.

The variations in animal mating are far too extensive for this book, but, suffice to say, while many use the scent of their droppings and urine, others perform some amazing rituals in order to attract the female. In some cases, such as the wolf, the pack is deliberately kept to a size sufficient to survive depending on the available food sources.

In a wolf pack, the alpha male is dominant but during the mating season, between January and April, the female alpha can take over dictating her wishes and deciding where the pack will locate their den. The rest of the pack will hunt and return with food for the laborious female and her pups. In a wolf pack, it is normally only the alpha male and female that reproduces to avoid the pack becoming too large. However, it is not unknown for a pack to have two dominant females. The pair will normally leave the pack for a few days during the actual mating to avoid any disruption from any other males.

ENVIRONMENTAL DISTURBANCE

All animals leave behind some disturbance of the natural habitat in which they reside. This includes territorial markings (the flattened grass and places where they have spent the night). In many cases, some animals dig burrows which leave a clear indication upon the earth, while others will actually build a bedding area. Feeding—especially herbivores—will leave behind broken and split branches in their forage for seasonal fruit and berries.

Environmental disturbance, which is clearly illustrated by the sleeping area of a Roe deer. Roe deer like to remove the leaves and sleep on the earth.

NATURAL HABITAT

There are many reasons for tracking animals: environmental, study of animal behavior and, of course, hunting for food. No matter the reason, in order to locate and track any animal, it is important to know what species live in what type of country in any given environment.

When hunting for food, game has to be of a certain size before we pursue it. Likewise, not all wild food tastes good; wild Canadian geese, a big bird available in vast numbers tastes really horrible.

Some animals are easier to locate than others, as some of which are extremely shy. Many of the larger animal groups—such as deer—will defend themselves if you get too close and it is only the sound of the gun that will panic them into a fleeting retreat. Bears are also animals to give a wide berth, simply because they are aggressive.

Lots of animals will openly show themselves if the conditions are right; a bright summer morning will reveal a lot of rabbits in certain countries, while ducks feel protected once upon the water. At other time you are best looking for animal habitats during winter when the leaves have fallen and it's easier to see through the forest.

The following list indicates most of the animals and their attributes. While the list is not exhaustive, it provides many of more common animals found in the wild. Where possible I have provided images of the animal and information that may prove useful to the tracker. To help distinguish between the various animals, I have made a small list of some of the more popular species (SH stands for Shoulder Height).

RED DEER

SH (Shoulder Height) up to 140cm
Young calves are spotted, although the adult red deer has a reddish/brown coat in the summer and a thicker brown/gray coat in the winter. They live in herds in open country and woods. The stag can be heard roaring during the October rut. Their diet includes grass, fruit, young shoots, and leaves of deciduous trees and shrubs. They will also eat bark.

Fallow Deer

SH about 100cm

Fallow deer fawns are born in June and their coats are spotted. The adult fallow deer coat is usually yellow above with white spots and yellowish/white below. During the winter, their coats are grayer with less obvious spotting. Their rut is in October, when the buck can be heard making belching grunts. They have palmate (flattened) antlers. Fallow deer prefer open deciduous woodland and tend to live in herds in parks and woods. Their diet includes grass, herbs, acorns and berries. They will also eat bark during the winter.

Muntjac (Barking Deer)

SH about 50cm

The muntjac either lives in pairs or alone. They have dark red coats marked with white on their rump—this is only visible when they raise their tails. Males have plain antlers pointing backwards and females have tufts of hair in place of antlers. Although the muntjac is a quiet animal, they will bark if startled. Their diet includes wild herbs, grass, shrubs, and shoots. They will also strip bark and raid crops. The muntjac is native to and remains widely in Southern Asia.

Female Roe Deer with her babies.

Roe Deer

SH about 75cm

They have a smooth red/brown coat during the summer and a long gray coat in the winter. They have three tines (the points that stick out off the main stem of the antler) on each antler. Roe deer prefer to live in open deciduous and coniferous woodland and are very common in Northern Europe. They live in small groups, although they can live alone. Their rut is in July/August. Their diet includes leaves, shrubs, herbs, grass, berries, and nuts.

Elk/Moose

SH about 1.8–2m

The elk or moose is the largest European deer in North and East Europe. Their coats are a gray/brown to black color, they have a beard and the males have large palmate antlers. They live mainly solitary lives, but can live in small groups. The elk/moose live in open forest, especially close to water (they are very good swimmers). Marshlands make an ideal setting for them in the summer, although they often move to drier land in the winter months. Their diet includes leaves, young shoots, water plants, grass, and moss. The elk/moose cannot be found living wild in Britain.

Sika Deer

SH 80–90cm

Their coats are brown and spotted in the summer, during the winter their spotting becomes less obvious and they turn a darker shade of brown. They have a distinctive white patch (with a black outline) on their rump. The sika deer rarely has more than four times on each antler. They are mainly nocturnal and live in small groups in deciduous and coniferous woodland. They were introduced by Eastern Asia to many parks in Europe, where escapes have meant that they can now be found in a number of countries, including Britain. Their rut is in October when the stag can be heard making short, screaming, grunts. Their diet includes shrub shoots, grass, and they will graze on young trees.

Wild Goat

SH about 75cm

They have horns with a simple, gentle curve that can grow up to 1m long. They tend to have a shaggy coat in the winter months and live in flocks on open hillsides, you will not find them roaming wild in Britain as they are restricted to other parts of Europe (i.e, some Greek Islands).

Wild Goat

Wild goats live in small herds. Their diet includes grass, leaves, lichen, and moss.

CHAMOIS

SH about 75–80cm

Both the males and the females have horns. They are goat-like with a light brown summer coat. They have a longer coat in the winter that can be nearly black. The females and young live in groups and the males tend to live a solitary life. They live in coniferous and deciduous woodlands or mountain pastures. This goat-antelope species is native to the European Alpine mountainous areas; they are also found in Romania, Turkey, and the Caucasus. Their diet includes grass, herbs in the summer, berries, and buds.

WILD BOAR

SH up to 100cm

The adult wild boar has a dense dark colored coat, although youngsters are striped. They have a long snout and the male has a large tusk, which can grow to around 30cm long. The male lives a solitary life, except in the rut. They are native to the deciduous woodlands in the woods and marshes of Northern Europe but can also be found through regions of Spain. They can also be found in Central

Wild Boar

America, Asia, America, and Australia where they have been artificially introduced. Their diet includes roots, acorns, bulbs, and fallen fruit such as acorns.

WOLF

SH about 1.2m

The wolf is a dog-like animal, with a broad head and chest. They have a bushy tail that hangs downwards and usually have a brindle coat, although this can vary. They are nocturnal and live in deciduous forests and tundra in remote areas of North America, Eastern Europe, and Africa. They live in small packs of between five and twelve, but through the winter months they join together to make large hunting packs. Their diet includes deer and small mammals.

BROWN BEAR

SH up to 2.5m

They are heavily built animals weighing up to 250kg (550lb). They usually live on their own and are nocturnal, although they are sometimes up and around in more remote areas. They can be found living in remote areas of Europe (not in Britain) and North America. During the winter they hibernate under ground. Their diet includes small mammals and they also dig for roots, bulbs, and insects. They are usually nocturnal and are active all year round. Their natural diet includes young deer, small mammals, birds, poultry, grass, and fruit.

ARCTIC FOX

SH up to 65cm

Unlike the red fox, the arctic fox is not nocturnal. They are a grayish/brown or a blue/gray color during the summer, turning white in the winter. They live in small groups and are active all year round. The arctic fox's preferred habitat is high mountains or tundra and they can be found in the Northern Hemisphere. During the winter months they may move into woodland. Their diet includes voles, lemmings, and birds.

EUROPEAN MINK

SH 35–40cm

The European mink has a coat that is dark brown above, apart from above and below the mouth which has white markings. They live solitary lives and the male is larger than the female. They can be either nocturnal or crepuscular (up and about at dawn and dusk). European minks are able swimmers with partially webbed feet and they tend to live on riverbanks and in marshlands. Their diet consists of small mammals, water voles, rodents, and fish.

OTTER

SH up to 80cm

Their coats are brown above; paler below. They have a long body, short legs, and webbed feet. Their ideal habitat is by lakes and rivers where they live solitary, nocturnal lives. Otters come in a wide variety of the species live throughout North America, Europe, and Asia. Their diet includes fish, frogs, and waterfowl.

BEAVER

SH up to 100cm

Famous for building dams in lakes and rivers, they also make 'lodge' systems which they live and breed in. They have a brown coat and webbed feet. Beavers can be found all over North America. They live in large family groups and are nocturnal animals. Their diet includes tree bark and water plants. Beavers are also found throughout Europe and have been introduced in Wales.

MUSKRAT

SH up to about 40cm

The muskrat is dark brown above and a creamy color below, with a tail that is flattened on each side. They either live alone or in small groups. They are diurnal and active all year round. The muskrat's burrow tends to be made in banks by freshwater and marshes. They also make breeding lodges in reed beds. They are native to North America can be introduced to parts of central Europe, Asia, and South America. Their diet includes water plants.

RABBIT

SH up to about 45cm.

Rabbits can be found throughout North and South America, European and Asia, as well as Australia. Their habitat is normally an extensive burrowing system called warrens, which can be found on farmland or woodland, although they will also live in sand dunes. Their legs are not long enough to hold them completely off the ground (like hares) and they tend to thump the ground with their hind feet if they are alarmed. Rabbits are active all year round and are crepuscular (up and about at dawn and dusk). Their diet includes plants.

HARE

SH up to about 65cm

They have very long black-tipped ears and long legs. The brown hare tends to live a solitary life and can be found in much of North America, Europe, although not in Sweden, Norway, Iceland, or parts of Ireland. They are usually active at dusk and at night and their ideal habitat is farmland and open woodland. The hare is extremely quick and can outrun most of its predators. They rest above ground in hollows (shallow depressions in the ground); these are known as forms, and their diet includes plants.

Arctic Hare

SUMMARY

The study of animal behavior is a very expansive subject, but for the purpose of tracking it can be reduced to knowing in what environment the animal lives; what its eating habits are; and when and where it feeds. Once this is known and the area found, it is possible to look for clues that will indicate that the animal is actually habiting the area.

In the main, these clues will be in the form of an actual sighting, prints, or droppings. As the number of animals is so diverse, it takes a lot of skill to actually locate and track an animal. However, the larger the animal the easier it is; deer for example, on pushing their way through the brush, will leave behind hairs as well as disruption of the vegetation.

It is best to concentrate on one animal at a time and get to know all their behavior patterns. Simple research prior to any actual tracking will save you a lot of time and trouble and help you identity your quarry.

Chapter 8

URBAN TRACKING

While man has been using tracking techniques that go back thousands of years, modern-day man has also developed a more technical approach to locating humans—the enemy in particular. They say that *'war is the mother of invention.'* If this is true, then much of that invention has certainly drifted down to the civilian level. One good example is the global positioning system (GPS), which we all use at our leisure to guide us safely to our location. It is not just our cars that are fitted with GPS, but ships, aircrafts, and just about every type of transport is guided by the system. In addition, we have helicopters fitted with highly powerful night vision cameras that patrol the darkness in search of villains. While over our borders and war zones drones fly silently overhead mapping the terrain and searching for any possible intruder or enemy. In their own way, these are modern forms of human tracking; from the offender who has an ankle tag to the surveillance satellite high above our planet, we are constantly being watched.

The human race—or more accurately governments—seems to have an inbuilt desire to track in some shape or form. We walk down the street in a crowded city center and the cameras we no longer notice are constantly watching us; while a person considered a threat is electronically tagged and kept under close observation. Then there are mobile phones, and everyone who has one in their possession can be located to some degree; either by the GPS chip or triangulation of its position to the transmitters. You use your credit card at an ATM while on vacation and immediately the authorities know you are in that country, that city, and at that specific ATM location. Moreover, if you come to the attention of the security services for any reason, they have the power to look deep into your everyday life. They can track your every movement, both electronically and by deploying a surveillance team. They can intercept your mail, fax, E-mail, and telephone conversations. They will plant cameras and microphones covertly inside your home, plus one outside your front door as a trigger when you leave or enter.

On the modern battlefield, surveillance is a vital element of information gathering. They have everything from radar, long-range thermal cameras atop extendable poles, and small and large surveillance drones fly overhead carrying

Every time you use an ATM you are simply providing your location.

out thousands of missions every day. And it is not just the battlefield; more and more this military technology is being used for border surveillance and tracking terrorist locations within a city.

Tracking does not stop the moment you leave the forest, desert, or jungle; tracking, albeit a totally different form, continues in the towns and cities around the world. Your footprint may be lost, but visual identification as to your location is still highly possible.

A good question is: Why do we track civilians? The answer to that is fairly simple: We are constantly under threat from one radical group or another. While major full-scale conflicts still happen, as we saw in the last Gulf War, today's problem is closer to home and the need to protect ourselves from those that would harm our nations.

Author's Note: I have spent much of my military life in surveillance, from the early days in Northern Ireland to the streets of London. While many of the techniques have been refined, the basic methods remain the same and it all comes down to skill.

A good surveillance operator is known as a *'gray'* person. That is to say they mingle with people but no one ever takes any notice of them. Their personality appears nondescript, they have no outstanding physical features, and their dress is innocuous. They are deliberately trained to be Mr. and Mrs. Nobody, so insignificant that no one ever gives them a second glance.

Yet this is only an outward appearance, as the surveillance operator requires many skills. They must be patient, as

Surveillance operators look more like normal people and deliberately make themselves look 'gray' and unnoticed.

surveillance operations can go on for months, even years. They must be adaptable, as many targets act erratically. If the target is a professional spy, they will deliberately check to see if they are being followed and take evasive actions in order to throw off any unseen surveillance operation against them.

Surveillance operators must have confidence, not only in their own abilities, but also in those of their team members. They must have a good memory, good hearing, and excellent eyesight... but most of all, they must blend into the background and almost become invisible.

As with visual tracking, we also need a starting point; a place where we know the quarry has been or can be located. In the urban jungle, most quarry live in a dwelling of some description (a house, apartment, or even a hotel). The first step is to glean as much information about the quarry and this is generally done by locating the quarry's domicile. As you would expect the military have a word for this task, it's called 'Target Recognizance' or target recce for short.

TARGET RECCE

Once the target has been identified, an in-depth surveillance operation may well ensue. However, as with most things military, the planning phase must first take place and a number of procedures must be followed. One of the first objectives is to gather detailed information on the target's known place of residence or their place of employment depending on the type of operation being undertaken. This information comes from a wide range of sources and is normally the start of the suspect target's personal file. Take, for example, that a known suspect has been seen talking to an unknown person. That unknown person is followed back to a house. Simple inquires will reveal the person's name and address. This is the start point from which some of the following material will be gathered.

- Aerial photograph.
- Detailed planning layout of the building.
- Any police or criminal records.
- Any driving convictions.
- Establish place of employment.
- Establish target's vehicle make and registration.

Today aerial photographs can be downloaded over the internet, even in 3D. These provide enough detail for planning a covert operation against the target's dwelling. In

Today the 3D satellite imagery is so good you can view it from all angles.

most police or military operations, a special flight may be flown to get the very latest images. Planning applications showing detailed internal rooms are also freely available from the local town hall. Likewise, police records and convictions are easy for the intelligence operators to obtain. However, one of the best ways into a person's life is to do a full *target recce,* a military term for really close onsite reconnaissance.

This will normally involve a small team of four men: one driving the drop-off vehicle, one watchman, and two to make the entry. The skills used by the group will be fairly standard such as one of the two-man entry team will be an excellent locksmith, while the other may be proficient in inserting concealed video and audio bugs.

Before any of this is done, a full outside reconnaissance will be done to establish the ideal time to make the entry into the property. If the person is living alone, then this is fairly simple as you just need to keep the quarry under observation while away from his domicile. If there are others present, especially a wife or children, then the best time could be early morning when the wife is taking the children to school. It is not my intention to write out a full description of how a reconnaissance team operates; needless to say they will gain access, even when the domicile is occupied.

The team will do a systematic search of the property for any incriminating evidence and at the same time gather basic profile information on the quarry. Any computers or tablets lying around will be fitted with a backdoor program for access at a later date. Bank details, credit card slips, fingerprints, and anything useful will be gleaned for further analysis. In most cases, the team will not leave a mess behind if there is any way to avoid it, as making the search look like a robbery will only make the quarry suspicious.

In many cases, electronic bugs and triggers will be placed in various positions so that the domicile can be monitored and a trigger provided when the quarry is about to leave. This will give any surveillance team time to get into position ready to follow.

There are 4.2 million CCTV surveillance cameras in the UK, that's one to every fourteen people; most people in the UK are exposed to CCTV at least one hour every day.

WHAT IS SURVEILLANCE TRACKING?

Surveillance is a technique used to obtain information, make connections, produce new leads, and collate evidence in a similar way to visual tracking. Surveillance can be carried out by a number of methods, including but not limited to:

- Human, Visual, and Audio
- Electronic, Video, and Audio
- Aerial and Satellite Surveillance

Surveillance may be carried out in order to obtain evidence of a crime or to identify persons who have been indicated in subversive actions. Surveillance methods help establish a person's location and lead to association with others. However, the main usage of surveillance is carried out by police or intelligence agencies. Governments have long since learned that information gathered on the potential lethality and capabilities of another nation helping prepare for defense or attack. One major problem with military intelligence is the amount of information they collect. Having information is one thing, interpreting its full and true value is quite another.

The enormity of this quandary is highlighted by the 9/11 Al-Qaeda attacks on America. The information that an assault was about to take place was available from several sources, the interpretation and immediate action was lagging too far behind. Organizations such as the NSA (National Security Agency) or GCHQ (Government Communications Headquarters) should have detected some traffic indicating a possible attack. Perhaps the perpetrators developed a method of sending messages that cannot be detected? Whatever the reason, it proves that electronic surveillance—no matter how sophisticated—is not as effective without good analysis and the correct distribution of intelligence. In the same way a visual tracker can derive information from a footprint, the electronic analyst must be skilled enough to interpret the data provided.

In reality, surveillance is simply monitoring the activity of a person or persons, a place, or an object. In order to do this successfully, intelligence agents need to consider several factors about the target. For example, if the target is a person then he or she will most probably move around, either on foot or by vehicle. On the other hand, the target may well be a house in the country, in this case a static observation position (OP) would be set up. The various methods of surveillance consist of one or a combination of the following:

- Static Surveillance
- Foot Surveillance
- Mobile Surveillance
- Technical Surveillance

Surveillance operators also have their terminologies which indicate a number of actions the target vehicle is likely to carry out. This terminology also helps the surveillance team keep the target vehicle visual and thus avoid a lost contact. While they vary from country to country, here are a few examples of vehicle surveillance terminology with an explanation as to what each means.

STATIC SURVEILLANCE

This is basically a place from where a surveillance team can observe their quarry's. For example, many of the foreign embassies will have a static surveillance location directly opposite their main entrance so that the host country can monitor who is going and coming.

Static observation posts are also used extensively by the military during any urban conflict. A perfect example was in the struggle between the British Government and the IRA in Northern Ireland. A suspect could be placed under twenty-four-hour surveillance, normally by a dedicated team being placed in an observation location which was best suited to watch the targets main residence. This would be used to ascertain who went to and from the residence and also to trigger a mobile team should the target leave.

FOOT SURVEILLANCE TECHNIQUES

In general, targets are not followed once, but many times. In doing so, the surveillance operators build up a pattern of the targets general behavior. In such a case surveillance it will be termed 'loose' and the operators will remain at a safe distance to avoid being compromised. Surveillance on the target can be done in short stages until a number of known 'triggers' can be identified, i.e, at 5:05 p.m. they leave their place of employment Monday to Friday. 'Loose' surveillance is normally carried out against a target that is living in a fixed location for a given period of time.

If the target has recently arrived in the country or has suddenly come to the attention of the intelligence services, then the initial surveillance will be 'close.' This means having the target visual at all times during the surveillance. This form of surveillance requires the very best operators in order to establish some basic information about the target, such as housing, employment, and associates. Once these basics are known, *'loose'* surveillance techniques can be employed.

Both foot and mobile surveillance operations have three distinct phases: the trigger or pick-up, the follow, and the housing. Any operation will be based on the fact that you need a starting place—normally a location you know the target to be—or where he will be going. The surveillance aspect is to follow the identified target and finally place them in a known domicile (the target house, for example).

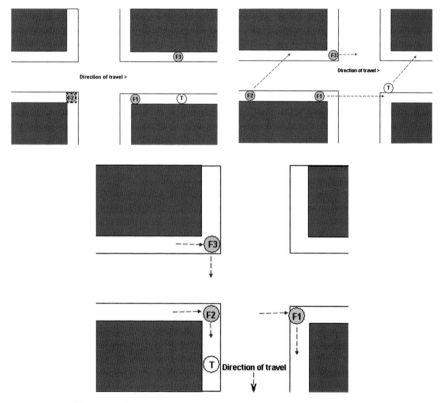

Foot surveillance team using a three-man follow technique.

Foot Surveillance Team

The basic foot surveillance team consists of a three-person unit. Their main objective is to keep at least two sets of *'eyeballs'* on the target at all times. An initial procedure for keeping a target under observation will be as follows:

- On target pick-up or *'trigger'* the first operator will remain behind the target.
- The second operator hangs back keeping the first operator in view.
- The third operator will walk on the opposite side of the street almost parallel with the target.
- On a linear follow, one and two may change places as and when necessary.
- If the target changes to the other side of the street, number three takes up the immediate follow with number two moving across the street as back-up. Number one will remain parallel to the target.
- On target turning left or right.
- Number one operator will go straight across the road and take up parallel position. Number two can choose to take up the lead while number three crosses the road to become back-up.

If the target is deemed to be of particular importance, then several foot surveillance teams will be deployed at the same time. It is possible during the follow for a target to adopt a mode of transport. For this reason, most surveillance is a combination of both foot and mobile.

If the target enters a telephone booth, number one should walk past and take up a location in front of the target. Number two should enter the adjacent booth—if there is one—enter money and make a real call (back to the office). Discreetly try to observe any actions the target makes and listen to his conversation if possible. (Never enter a phone box while carrying a mobile phone—this is a dead giveaway.)

If the target drops an item, it should be collected. However, be aware that this can also be a ploy on behalf of the target to see if he is being followed.

Always make a note of any person the target gives an article or money to. While the purchase of a newspaper may seem innocent, it is also a perfect way of passing a message.

Additionally, log all known purchases, as this provides an idea of how much money is being spent by the quarry.

Author's Note: It is interesting to note that the largest ever spy ring caught in the UK all started when a minor suspect, who worked at a secret factory making a new type of torpedo, was apprehended. The worker had been coherced and bribed into smuggling drawings of the new torpedo out of the factory. By chance, a surveillance operator noted that the worker in question led an expensive lifestyle and summed up the amount of money he spent. This eventually led to the discovery of a massive spy ring known as the Cambridge 5. Guy Burgess and Donald Maclean both fled Britain for the former Soviet Union when another member, Kim Philby, tipped them off. At the time, Philby was head of counter-intelligence for British Intelligence; he later became suspicious of being detected and he also defected to Russia.

Mobile surveillance involves using every trick in the book.

Mobile Surveillance Outline

Mobile surveillance is the use of vehicles, boats, or aircraft in order to follow a target that is also mobile in some mode of transport. This type of surveillance requires skillful driving, good observation, set procedures, and excellent communication. It also takes a lot of discipline on behalf of the driver, as surveillance often turns into a chase rather than a discreet follow.

The same basic principles that apply to foot surveillance also apply to vehicle surveillance. However, the basic principles of vehicle surveillances are more difficult because of the complications created by traffic congestion, restrictions imposed by traffic laws, and the greater possibility of the operation being discovered. Like foot surveillances, an individual's vehicle will be limited in its capability, where a team's vehicle acting together enhances the prospects for a successful operation.

The surveillance vehicle should accommodate either two or three people, making a foot follow possible in the event that the target goes foxtrot. Having at least two people in the vehicle also allows the driver to concentrate on his driving while the passenger is alert to the surroundings.

There are numerous techniques the driver can use to reduce the risk of detection, such as having the ability to switch off one of the headlights during a nighttime follow. This provides a different pattern to the target if he is watching in his rearview mirror. Likewise, a distinctive feature can be attributed to the target's car, such as smashing the tail light. To make this look natural, always do it when the vehicle is parked in a busy street or car park. Also be aware that this may tip the target off to the surveillance operation.

When the target vehicle is temporarily parked, one of the surveillance operators should go on foot while the other remains with the vehicle. If the target vehicle is parked for any length of time, the surveillance vehicles should move their position intermittently. Those remaining in the car should also sit in the passenger seat to make it appear they are waiting for someone.

Good area knowledge will avoid the necessity of constant map study, which means taking your eyes off the target. However, the introduction of on-board GPS tracking devices has, to some degree, alleviated this problem. But before any surveillance operation can begin, certain questions must be asked.

- Is mobile surveillance the best way of achieving your goal?
- Is the operational area well known?
- Type of operational area, urban or rural, i.e, will foot surveillance be required?
- Is the target's awareness level known?
- Is the target's vehicle known?
- The pick-up point or trigger.

The answer to these questions is normally self-evident. If the target is likely to go any distance, it may be better to employ a helicopter than to deploy six mobile surveillance vehicles. Likewise, the target may use a form of transport that can outrun or out-maneuver the surveillance vehicles (a motorbike for example).

Knowing the trigger for the target vehicle is also vital. In the normal course of events, all team vehicles will familiarize themselves with the streets around the trigger. A briefing is given before the operation begins, at which time all surveillance vehicles will be given their call signs, assigned start points, and time in position (known as the plot up). Once on the ground, one vehicle may decide to do a drive past or put an operator on foot in order to confirm the target's location. (Are the house lights on? Is the target's vehicle parked outside?) The pattern of the vehicles will be set as to trigger a follow, irrespective of which direction the target vehicle drives. As with foot surveillance, the actual trigger will come from either a static OP, foot operator or one of the surveillance vehicles. Communications are tested and all vehicles confirm *'in position.'*

Typical Terminology Used by Vehicle Surveillance Teams

Back-up, can you?: Current tail car (normally known as *'eyeball'*) request to back-up vehicle to ascertain whether handover is appropriate. Response is either yes or no or backup can at next junction.

Backup: Second vehicle in convoy.

Cancel my last: Ignore instruction or information just given.

Come through: Given after *'hang back'* to bring the convoy through.

Committed: Indicates that the target vehicle is committed to traveling on the motorway.

Contact, contact: Indicates that eyeball has been regained by one of the vehicles in the convoy, following the search procedure. The pick-up vehicle will also give location.

Convoy check: Request from eyeball to determine position of vehicles in convoy, to which all vehicles automatically respond in turn. Motorcyclists should respond without specifying their precise position, after Tail End Charlie has reported. When all correct, eyeball calls *"convoy complete."*

Down to you: Final transmission from eyeball handing over surveillance to another vehicle.

Eyeball regained: Indicates that the target vehicle is once more under surveillance.

Eyeball: Vehicle or officer having primary visual contact with the target and who is directing the operation for the time being.

GPS Vehicle Tracking

There are many devices that are fitted to vehicles in order to track their whereabouts. Literally thousands of companies now track their vehicles simply to adjust and optimize the best route and save on fuel or simply to know the location of the vehicle.

However, when a vehicle needs to be tracked covertly without the owner's knowledge, it is expedient to learn how to do this properly. Having been responsible for attaching a tracking device to many a vehicle, I found there is only one professional way and that is to be totally prepared beforehand so that you spend the minimum of time below the vehicle.

There are several ways to do this but it has always been my belief that you first need to gain access to a similar vehicle before approaching the real target vehicle. That is to say, if your target car is a BMW 5 Series, then you need to get your hands on a similar car. I always went to a car hire company and rented the car I required and then used it to ascertain where and how I would conceal my tracking device. Having a similar vehicle at your disposal for a few days allows you to check the best possible location for the tracking device where the optimum signal

Drone surveillance is growing at an alarming rate.

can be obtained. It will also help you find the relevant electrical wires so you can permanently power the device. Once you are happy with everything, you simply take your device and wire it into the target vehicle. This means you will know exactly where to place it and how to keep it powered—added to which you will spend the minimum amount of time on site.

Drone Surveillance

Almost every day we hear of some incident that has taken place around the world which involves the use of drones. The larger, armed drones track down and eliminate terrorists or enemy camps, while at the other end of the scale the civilian demand for small utility drones is growing at a phenomenal rate.

In late 2013, Amazon announced that they were testing small drones for use in the delivery of parcels. The drone called Octocopter could deliver packages weighing up to 2.3kg (5lb). However, this has yet to be proven as battery power is still a major problem—plus they will need approval from the FAA (Federal Aviation Administration). One thing is for sure: Amazon is serious about the project and with their financial pull it could be with us in as little as five years from now.

In the meantime most of the drones are being built and used by the military and, despite the growing opposition from the public, drone strikes continue. On October 28, 2013, an American Reaper drone launched two missiles at a Suzuki vehicle which was carrying two known members of Al-Qaeda. The attack took place near the town of Jilib in the south of Somalia. The vehicle was completely destroyed and the occupants killed.

Drones both large and small continue to be developed for an ever increasing number of uses. The *SQ-4 RECON* range is a new breed of Advanced Aerial Robots (Quadcopter) that has been developed specially for the military/police/homeland security/search and rescue/HASMAT operations. The operational performance and functionality allow the drone to be flown either with its own ground control station or controlled via an embedded secure website supported by any modern web-enabled device. This unique feature allows for the *SQ-4 RECON* to be flown virtually anywhere in the world, from anywhere. The drones innovative features cover such as 'Sense and Avoid' sonar, and two-way voice communications via a 'voice over internet' protocol.

The *SQ4-RECON* is fully autonomous, which means it can interact, avoid obstacles or humans, land, return to operator, hold position using GPS, inertial systems, or the sonar. It can also be controlled in manual mode where a human operator issues regular commands via a push button panel; functions cover all normal movement: forward, backwards, rotate or climb, etc. or it can be run in way-point mode. A human operate can talk through the drone over its WLAN structure in addition pre-recorded voice messages can be played when flying in autonomous mode. This function is designed for deployment in search and rescue operations/ crowd control/law enforcement, etc.

Drones will continue to grow in functionality and fly safely and in very near future. They will be as common a sight as the surveillance camera which most of us now take for granted.

UK Emergency Aviation

Once a police helicopter has located you, it's extremely difficult to shake them off as they are fitted with both day and night visual.

Airborne Surveillance

The use of helicopters in surveillance is a great asset as it allows the follow vehicles to hang back and avoid being detected by the target. While a helicopter is easy to spot, most can sit off the target by several miles and still keep track on the vehicle through the use of powerful cameras, most of which have day/night capability.

Helicopter surveillance has become popular with the police, providing an overt observation platform for many different operations, such as traffic control and police pursuit. Helicopters also have the advantage of speed and unrestricted progress while in the air, making them ideal for:

- Surveillance
- Aerial photography
- Aerial reconnaissance
- Electronic tracking
- Communications relay
- Insertion and extraction of agents

Rapid Back-Up

In addition to helicopters and light aircraft, the use of unmanned drones for military surveillance has long been established, but new smaller models have been perfected for civilian surveillance use, some no larger than an insect (see reference to Chapter on Technical Surveillance earlier in this chapter).

SURVEILLANCE CHECKLIST

No matter what form the surveillance tracking takes, be it a static observation post, foot, or vehicle surveillance, there is a standard procedure to reduce the possibility of the quarry becoming aware. This procedure requires defining the operation objectives and the best way to obtain the accurate results without being detected.

- Research all available information on the target.
- Obtain photographs and physical descriptions.
- Establish license numbers and makes of vehicles used by target.
- List known associates.
- Is target likely to be armed?
- Detailed sketch of target premises or aerial photograph.
- Define points of entry and exit.
- Detailed street map of the target premises and surrounding areas.
- Establish codes for target, associates, locations, and alternate plans, etc.

MANNING REQUIREMENTS

There is also a need to outline the number of surveillance operators required for the various types of surveillance required, i.e, static, mobile, or technical. If the team is to go mobile, then it must calculate the minimum number of vehicles required, including backup. The commander must identify specific operators for individual tasks, such as Methods of Entry (MOE).

In many cases, it will be best to deploy a mixture of male and female operators taking into account any ethnic majorities in the area of operation. Dress, language, and role all play a major part in being undetected. If you need to loiter, then the surveillance operator must have a legitimate cover, one that goes unnoticed by the average person (road sweeper for example). There must be a prepared list of equipment and this must be working and tested, it is no good being onsite with useless equipment:

- Check and test all radio equipment (vehicle and personal) include spare batteries.
- Issue adequate funds (including change) for telephone calls, parking, meals, etc.
- Request all forms of technical equipment required, cameras, binoculars, etc.
- Consider a change of clothing or disguise in vehicle.
- Recovery procedure in the event of a breakdown or accident.
- Field tests all communications with base station.
- Install repeaters in areas of poor communications if required.

OPERATIONAL BRIEFING

A clear set of operational orders should be prepared so that the team knows exactly what is expected of them. During the brief, the commander should reinforce the problems arising from compromise. Emphasize the need for safe and discreet driving practice. Distribute all available data, photographs, telephone numbers, etc. and designate radio channels to be used while reinforcing proper radio procedures.

It is always best to do a dry run so that the team can familiarize small administrative points, such as making sure all the vehicles are fully fuelled prior to any surveillance (nothing worse than having to stop and refuel midway through a follow).

There should always be a post operational debrief which allows all those who took part in the operation to have their say. This should include route taken, any deliberate stops made by the target, photographs, and video taken. Mobile surveillance has a habit of going wrong and the debrief must deal with any points of possible compromise, with any solutions being discussed

COUNTER-SURVEILLANCE MEASURES

If you believe that you might be under surveillance or are being tracked, it is equally as important to spot the tell-tale signs. Those that believe they are under surveillance or simply suspicious to the fact may choose to purchase and use specialist equipment to confirm this. Always remember: anything you do, write, or say can be monitored by a myriad of technical devices.

If you are a subversive working in a foreign country, you must assume that you are under surveillance. Everyone can be bugged for any amount of reasons. There is a general list of counter-measures most subversives will look out for if they suspect that some form of technical surveillance will be targeted against them.

When you have detected mobile or static surveillance in the past, yet recently you are convinced that they have stopped watching you, they have planted and are using technical surveillance instead.

You detect all is not right in your home. The furniture seems to have moved, your personal effects are not where you left them.

Your home or office has been burgled, but nothing of significance has been stolen. That could mean a professional team has entered your premises and implanted a number of technical devices. Check for loose plaster work or plaster crumbs. Check all electrical fittings, including your phone. Check the walls and ceilings for any telltale signs or bulges.

Open and check any fixed items such as fire alarms, plug sockets, light fittings, and wall clocks.

The door locks are not working as smoothly as they have done for years . . . a good indicator that someone has been using lock-picks to gain entry. Install a dead bolt-type locking system, heavy enough to stop the average locksmith. Check the external door frame for indentations. This could mean that a hydraulic jack has been used to spread the door frame and release the locks and bolts from their housing.

Your phone is making odd noises; it rings and there is no one there; you can hear a tone when the phone is on the hook. All these indicate a telephone tap.

The television, car radio or AM/FM radio develops strange interference. This could mean that the unit has been tampered with and a hidden wireless microphone implanted. You might well be picking up static from a device near to the television or radio.

Be wary of any sales person offering you a free gift, such as a pen, cuddly toy, or clock radio. These can all contain hidden audio/visual devices with a wireless transmitter. Take notice of any van type vehicle that has suddenly started to appear in your street. These are usually disguised as utility or trade vehicles. Check the vehicles with a walk-past. If you cannot see clearly into the whole vehicle you must suspect that it is a technical surveillance vehicle. Use a stethoscope pressed against the window pane in order to detect any microwave *'buzzing.'* Check for any vehicles in line of sight of the window.

Never allow anyone to enter your premises without good reason. Telephone companies and electrical engineers do not just *'turn up.'* Check the identity of anyone you are not sure of. Watch them while they are working if you are suspicious.

Detecting Counter-Surveillance

It is important for the surveillance operators to recognize the signs of counter-surveillance. This helps to identify whether the target is actively engaged in unlawful activities or normal social behavior. A known target who is about to meet his handler or agent will almost certainly carry out some counter-surveillance techniques. When doing so, the target will watch to see who reacts unnaturally or is taken by surprise. He will observe any person who suddenly changes direction or looks to be giving a signal to another person.

- Stopping, turning and looking at anyone to their rear.
- Making a sudden change of course or reversing their course.
- Walking slowly then speeding up suddenly.
- Turning a corner and stopping to see who comes around the corner.
- Walking around the block.
- Going into a building (pub) and immediately exiting via another door.
- Checking constantly in the shop window reflection.
- Waiting to the last minute to step onto a bus or underground train.
- Getting off at the next stop, waiting and catching the next bus or train.
- Deliberately dropping something to see if anyone picks it up.
- Changing their appearance or clothing.

A quick change of identity can be achieved by adding accessories or changing your clothes.

SUMMARY

While this book is primarily about visual tracking, this chapter highlights the vast changes made in technical tracking. As the bulk of the population now live in villages, towns, and cities, and the factors of traditional tracking become unobtainable, it is incumbent on society to be able to track people when the need arises.

Incredible, it would seem, that the more sophisticated society becomes, the easier it is for people to be tracked. Our individual footprint as we go about our daily life is recorded by so many different methods, tracking has become almost automatic. Our movements—and proof of our movements—is written in every bank statement, recorded on CCTV, or shown on our mobile phone activity. If that's not bad enough, when governments want to track you in an urban situation they have vast resources to call on, and once all the 'bugs' and surveillance teams have been deployed, finding your exact location really is easier than visual tracking.

Chapter 9

ADDITIONAL TRACKING SKILLS

While there are many skills required to be a good visual tracker, there are two that need a specific mention in detail; these are navigation and camouflage. They are at the very heart of any good tracker's skills and equal to anything else you might learn. If you cannot navigate, you will get lost and be unable to inform others as to your location. If you do not obey the basic laws of camouflage, the quarry will detect your presence and flee.

The first concern to any visual tracker is the terrain. It can be difficult to track if you don't have a fundamental knowledge of the land over which you must work. Study every available map, air photos, and get a detailed briefing from the local police, military officials, and population. Talk to local farmers, natives, pretty much anyone who has been in the area in which you will be operating.

A military tracking team is just the first stage in any 'follow up' operation, and while they may be in sufficient force to defend themselves, they will more than likely have a support group on standby ready to helicopter into the area should they locate the enemy (see Chapter Five).

There are lots of ways of navigating; having a basic understanding of map reading, such as reading a map and using a compass. More modern devices such as GPS are great for pre-set routes, but not so clever when used for tracking. A professional tracker will always keep their direction of travel and distance at the back of their mind and have the ability to use natural indicators such as the sun, moon, and stars when visible as sign posts in a similar way to the tracking signs left behind by the quarry. Additionally, they will understand the terrain over which they travel and have good knowledge of the average speed they are traveling while pursuing the quarry. Both natural and man-made features such as rivers, roads, and prominent tracks will act to confirm location in the tracker's mind. Above all, the professional tracker will have a built-in sense of direction and rough current location purely from previous experience and local knowledge.

NAVIGATION

In essence, there are four basic ways in which one can navigate: using a map and compass; a Global Positioning

System (GPS); by using the sun, moon, and stars; and local knowledge of the terrain. The correct use of a map and compass is a basic skill that every tracker should learn until they are fully competent in navigational techniques. Other navigational skills, not dependent on a map and compass, can also be studied and are extremely useful in situations where the terrain is unknown. These basic skills will prove useful if your compass or GPS gets lost or damaged.

Maps

A map can be an essential tool to help you plan and follow a route through most terrain. Sometimes if the area is heavily forested, it may be difficult to see the actually route; but with the aid of a good map you should be able to follow it. Although there are a number of good local maps, the best type to obtain is one drawn to military standard.

Map & Compass, the basis of all good navigation.

The most British OS maps are similar to military maps (printed by the same government office) and these come in a wide variety of scales. The 1:25,000 scale Pathfinder map is ideal for navigation across most of the UK (US equivalent Geological Survey maps). No matter what type of map you choose, you will need to look at it and understand what you are seeing. That is to say, the image on the flat paper map must suddenly jump into a 3D image in your mind. With practice you can do this as the contour lines around the hills and valleys present a shape that can be translated into a mental picture.

Author's Note: If you have never done any map reading before and this is your first time, try to locate an area with some hills and valleys. Get yourself into a good vantage point and orientate the map to the ground, i.e, turn the map so that it corresponds to the view you are looking at. Now look at a prominent feature such as a mountain and locate the same feature on the map. The contour lines will be close together where the mountainside is steep or wider apart where the mountainside has a smoother slope. Try to visualize something else from the map and then look at the actual feature on the ground and see if it is how you imagined it. It may take some time, but this is an excellent way of training your mind to understand what the terrain is actually like from just looking at a map.

COMPASS

A compass is a precision instrument used for navigation. They come in a variety of shapes and sizes, but all work on the principle of a magnetized needle continually pointing north. A 'Silva' type compass is the most popular with civilians and military personnel as it lends itself to the map in a variety of ways. Most models are made of clear plastic with the compass housing containing the magnetic needle offset to the left side. The base of the compass has a magnifying glass and is etched with a variety of scales and a number of roamers to calculate grid references. The rim of the compass housing—which can be rotated—is marked with segments showing degrees, mills or both, while printed on the base is an arrow and orienteering lines.

Always remember that any compass works on the magnetic attraction situated close to the North Pole, localized power supplies, or heavy metal objects can pull the needle from its correct course. Most compasses dampen the movement of the needle by filling the compass housing with a liquid. This sometimes produces a bubble, providing the bubble is not large it should not affect the compass operation.

Setting a Map by Inspection (Orientation)

Look for an obvious and permanent landmark; for example a river, road, or a mountain. Find the feature on the map and then simply align the map to the landmark. The map is now set to conform to the surrounding features.

Setting a Map by Compass

Pick a North-South grid line on your map and lay a flat edge of the compass along it. Then, holding the map and compass together, turn both together until the compass needle points North. The map is now set to conform to the surrounding features.

Finding a Grid Reference

When you look at the map, you will see that it is covered in horizontal and vertical light blue lines. These are called grid lines and on a 1:50,000 land ranger map are 1km apart. The vertical lines are called eastings: these are always given first. The horizontal lines are called northings: these are given after the eastings. Each grid square is defined by the numbers straddling the left grid line of the easting and the center bottom of the northing. For example, the illustrated grid square reads 1729.

Usually, a grid reference contains six figures. This is worked out in the following way: the grid square is mentally divided up into tenths; for example, halfway up or across a square would be '5.' This reference point is then added after

To set a map by inspection it is a simple manner of turning the map to correspond with the actual terrain. Choosing two prominent features such as hilltops will do the job.

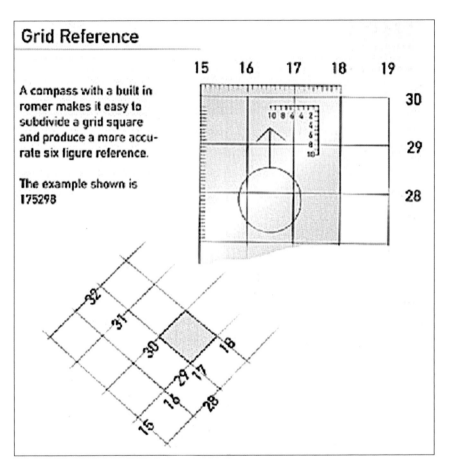

Grid Reference

A compass with a built in romer makes it easy to subdivide a grid square and produce a more accurate six figure reference.

The example shown is 175298

Grid reference of a built in romer.

the relevant easting or northing figure. To gauge the tenths accurately, use the roamer on the compass, or a protractor. The grid reference of shown would be calculated as 175298.

Taking a Compass Bearing from the Map

Once you have established where you are and where you wish to go, work out your route. Study the map and the distance. Plot the most logical route to your objective, taking into account the terrain and any obstacles. Divide your route up into legs, finishing each leg close to a prominent feature if possible, i.e, a road bridge, trig point, or even the corner of a forest area. Take a bearing from where you are (call this point A) to the feature at the end of your first leg (call this point B). Place one edge of the compass along the line adjoining A and B, making sure that the direction of travel arrow is pointing in the way you want to go. Hold the compass plate firmly in position and rotate the compass dial so that the lines engraved in the dial base are parallel to the North-South grid lines on the map. Finally, read off the bearing next to the line of the march arrow on the compass housing. To walk on this bearing, simply keep the magnetic arrow pointing north over the etched arrow in the base and follow the line of the march arrow.

The bearing gives the direction to a certain point. It can be defined as the number of degrees in an angle measured clockwise from a fixed northern gridline (easting). The bearing for North is always zero. Most compasses have scales of 360 degrees, or more normally they are shown in Mils with 6400 mils in a full circle. Some compasses have both scales.

Keeping on Course

Three factors will determine which route you take: the weather, the time of day, and what the terrain is like between you and your final destination. In good visibility, select features that are both prominent on your map and visible to the eye. Once you have taken a bearing, choose a feature along the direction of travel and head towards it. This saves you constantly looking at your compass. It will also help keep you on course if the terrain pushes you off track, i.e, you are forced to contour or avoid some obstacle. Success in reaching your final goal is having confidence in your route selection and not becoming a slave to your compass. Mistakes in poor visibility can be avoided if you consult the map every time you meet a prominent feature. As previously mentioned, careful study of the map should provide you with a mental picture of the ground relief, which in turn, will warn you of any obstacles, such as river or marshland, etc.

> **Author's Note:** There is a tendency during fog or poor visibility to wander downhill when you are contouring. Every hundred meters or so take a few steps uphill to compensate for this. Do not forget that you will move slower in poor visibility.

Putting a Compass Bearing on the Map

If you become disorientated, here is a simple way to pinpoint your position. This is done by locating a couple of landmarks that can also be identified on the map. Point the compass at the first landmark and, holding it steady, turn the housing until the orientating arrow is aligned with the magnetic needle. Now read off the bearing to the landmark. For example, say the bearing was 5700m—calculate if you wish the magnetic variation, which we will say is 40m, and subtract. This leaves us with a revised bearing of 5660, for which the compass dial can be adjusted. Placing the top right-hand edge of the compass against the landmark and pivot the whole compass until the orienteering lines in the base of the housing are running parallel to the eastings. Draw a line.

Find another landmark and repeat the whole procedure. For example, the second bearing is 0650m, 0610m after adjustment for the magnetic variation. Draw another line as above. Your position is marked where the two lines cross.

GPS is one of the world's best navigational systems, but it can be switched off.

GPS (Global Positioning System)

Today we all take GPS for granted and use it to locate an address or take us from A to B while traveling. This high-tech method of navigation is worthy of a mention, as it has started to replace the traditional compass, although not entirely. (Try using your Satnav in the jungle and see how far you get!)

Developed by the United States Department of Defense, the GPS system consists of numerous military satellites which orbit the Earth, continually giving out the time and their position. Receiver units on the Earth pick up this information. These units, known as GPS, have advanced at a phenomenal rate and although designed primarily for the military, Satnav is now available on most mobile phones. The GPS unit is able to receive and assimilate information from several satellites, converting it into a recognizable position and altitude at any point on the earth's surface. Receiving units vary, as do their accuracy, but most are good to within around 3 to 10 meters. Most Satnav mobile phone apps will also display mapping in a variety of forms allowing both location and surrounding terrain to be seen.

How it Works

The GPS receiver unit searches for and then locks onto any satellite signals. The more signals you receive, the greater the accuracy, while a minimum of four is sufficient, six satellites will give you a better fix. The information received is then collated into a usable form; for example, a grid reference, height above sea level, or a Longitude and Latitude. Individual requirements for use either on land or at sea can be programmed into the unit.

By measuring your position in relation to a number of known objects, i.e, the satellites, the receiver is able to calculate your position. This is called satellite ranging. It is also able to update your position, speed, and track while you are on the move and can pinpoint future way-points, thereby taking away the need for recognizable landmarks.

Author's Note: Despite its excellent qualities, the GPS system can be shut down and, if the signal from the satellite is blocked by overhead foliage, you can lose GPS coverage altogether. Additionally (and most importantly), the use of continuous GPS and mapping will diminish your battery very quickly, so do not forget your compass.

FINDING DIRECTION WITHOUT A COMPASS

Compasses may be the easiest and most convenient method of finding a direction, but what if you are without one? Many people wander off using just a map, which is fine until you get lost or the weather and fog disorientate you. Luckily, there are a number of other methods to find direction—all that is needed is a bit of intelligence.

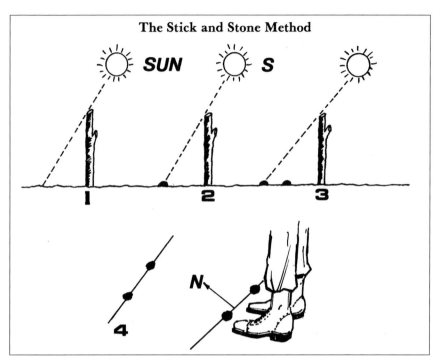

The Stick and Stone Method

Using a stick and stone is a simple way to find direction.

Stick and Stone Method

- On a sunny day find or cut a stick about one meter long and push it upright into some level ground. The stick will cast a shadow.
- Using a small stone, mark the end of the shadow as accurately as possible.
- After fifteen to twenty minutes the shadow will have moved. Using a second small stone, mark the tip of the new shadow.
- On the earth, draw a straight line that runs through both stones. **This is your East-West line.**
- Put your left foot close to the first stone, and your right foot to the second stone.
- You are now facing North.

Author's Note: The accuracy of this method depends on how level the ground is, how well the ends of the shadows are marked, and how much care was taken in placing the toes at the line. A North-South indicator can be produced if a line is drawn at right angles to your East-West line. Any other direction can be simply calculated from these cardinal points.

Using a Watch

Britain, which lies in the northern hemisphere, allows us to use an analogue watch in order to set direction.

- Check that your watch is accurately set to local time.
- Point the hour hand at the sun.
- Using a thin twig, cast a shadow along the hour hand through the central pivot.
- Bisect the angle between the hour hand and the twelve o'clock position.
- This line will be pointing due south, north being furthest from the sun.

By Night

Navigation by the stars has been used for centuries and is still employed in map making. Learning about the stars is beneficial in itself, but this knowledge comes into its own during survival navigation. Bright stars that seem to be grouped together in a pattern are called constellations; the shapes of these constellations and their relationship to each other do not alter. Because of the earth's rotation, the whole of the night sky appears to revolve around one central point and using this knowledge can help you to find directions.

By night, knowing where to find the North Star is extremely helpful for navigation

The North Star

In the Northern Hemisphere, a faint star called Polaris, the Pole, or the North Star marks the central point. Because of its position, it always appears to remain in the same place—above the North Pole. As long as Polaris can be seen, the direction of 'True North' can be found.

To find Polaris, first locate the constellation known as 'The Plough' or 'The Big Dipper.' The two stars furthest from the 'handle' always point towards Polaris. Take the distance between the two stars and then follow the line straight for about six times the distance. At this point you will see the Pole Star.

If you are unsure which way to look or wish to confirm that you have found Polaris, look for another constellation called Cassiopeia. The five stars that make up this constellation are patterned in the shape of a slightly squashed 'W.' It is positioned almost opposite the Plough and Polaris can be found midway between them. As long as the sky is clear, the Plough, Cassiopeia, and Polaris remain visible in the sky all night when seen from any country north of 40 degrees North latitude.

Check your equipment and pay special attention to camouflage and noise. Make sure you are only carrying the essentials and check the rest of the team for the same.

Tracking is not just about following the spoor, you may be entering enemy territory or carrying out cross border operations. In this case you will need to survive as well as track. This means locating drinkable water and a safe place to harbor up. Also check out the local weather for the next few days, as inclement weather can seriously affect any tracking operation.

While tracking any quarry there is a need to know where you are at all times; this is particularly important if you are a military tracking unit. Any tracker that follows a quarry without knowing their rough location is simply asking to get lost—in some environments this could be extremely dangerous. A professional tracker will know (roughly within half a mile or so) exactly where they are at all times and have the ability to withdraw should it become necessary.

CAMOUFLAGE

Camouflage is used by both the military tracker and the civilian hunter; except to the military tracker it can mean the difference between life and death. Camouflage works two ways: you try to make yourself undetectable, while your quarry or enemy will do the same. Camouflage, concealment, light, and noise discipline are vital to any successful tracker.

To camouflage successfully, you must constantly consider the terrain you are moving over and adapt your concealment accordingly. This can be achieved by studying the main factors that govern camouflage and concealment. The main recognition factors are based on object movement, contrast, or pattern change to the normal background. Most soldiers and hunters will know this but, if you're new to visual tracking, you will need to understand why camouflage is so important.

Good camouflage, but the soldier on the left could have concealed the white patch on his rifle.

Why We Are Seen

We see and are seen by the same factors. In addition, noise attracts attention. You may cover yourself with camouflage, remain hidden and silent, but if you do any of the causes listed below, you may expose yourself. When the enemy is watching, it takes very little to give away your position. Ask any good sniper who has lain in wait for maybe days just to spot one single sign of the enemy and they will tell you 'movement' is the major factor when it comes to giving away your position. To fully understand how a person is spotted we need to look at all the factors to which a person gives themselves away.

Movement will
give your
position away.

Movement

In order to distance themselves from a possible tracker, the quarry must move. For me, movement will attract attention much quicker than most of the other factors with the exception of sound. While trees and plants move and grass sways, there is a definable difference when humans or animals move; this is even more pronounced against a sedentary background. Rapid movement is easier to detect than slow movement; that is to say a man crawling on the ground will not present a visual picture as clearly as a man running.

Shape

No matter how good military camouflage against the natural background is, it is always going to present some features with hard, angular lines. Even a camouflage net—when erected—takes on a shape with smooth curves between support points. The threat can easily see objects if they are silhouetted, but they can also see objects against any background unless you take care to conceal them.

Shine

Anything that shines is a perfect eyecatcher. This may simply be the sun on a rain drop or the sun reflecting off a vehicle windscreen. No matter the cause, shine is easily detected. It is also one of the attributes that a good tracking team should take all measures to eliminate within themselves by reducing any personal objects that may

Shine, such as sunlight on metal or glass can be seen miles away in open country or desert terrain.

cause shine. An example of this is a mess kit, watch face, binoculars, or inappropriate use of a signal mirror.

Texture

When the surface of the ground is disrupted by footprints or tire tracks, the texture of the ground appears to change color. This is simply due to the light changes made by the disruption making a smooth surface look darker. Anything that changes the texture will result in a visual sign; footprints along otherwise flat sandy seashore are a perfect example of texture change. Flat uniform areas in any landscape will always stand out and provide visual sign of any disruptive texture, yet at very long ranges and in poor light all colors tend to merge into an even tone.

Silhouette

To see a tree silhouetted against a clear skyline around dusk is something most of us have experienced. When you are trying to remain hidden and camouflaged, the last thing you want to do is silhouette yourself in a similar way. A silhouette is a clear dark shape against a light background.

Shadow

Shadow can be divided into two types: cast and contained. Cast shadow is the type we are most familiar with and is the silhouette of an object projected against its background. Contained shadows are the dark pools of shadow formed in permanently shaded areas. Examples of contained shadows are those under the track guards of armored fighting vehicles (AFVs) inside a slit trench, inside an open cupola, or under a vehicle. These shadows show up much darker than their surroundings and are easily detected by the enemy.

Military Equipment

Modern armies strive hard to mimic the natural terrain by providing the soldier with camouflage clothing and equipment. In my lifetime, I have worn and seen at least a hundred types of military camouflage and while some have been good, very few have actually done the job they are intended to do—mask the soldier. In some cases, the camouflage pattern has done more to highlight the soldier's position rather than conceal it. The same goes for the

soldier's equipment, "a rifle is a rifle," there is no denying the shape; the same goes for their water bottle and magazine pouches—to the trained eye they are easy to detect.

CONCEALMENT

There are various methods of concealment, all of which involve not being seen or detected. Traditionally, the camouflage net has been deployed for both soldier and their vehicles. They cover mines with earth or grass that resemble the surrounding area and they make hides in which they live or wait in ambush.

The true art of concealment is blending in with your background, becoming a part of the scenery. Where the object they wish to conceal is angular, such as a tank, then efforts are made to disrupt the shape of the vehicle so it is not so recognizable.

Disguise is also a way of concealment. For example, making an artillery gun look like a military truck is a form of disguise. As is the Special Forces soldier who dresses in civilian clothing to mingle with the local population in Afghanistan. Disguise is the application of materials to hide the true identity of a person or object. The purpose of disguise is to change the appearance of an object to look like something else.

While most military equipment is camouflaged to suit the terrain, it is the shape that gives it away.

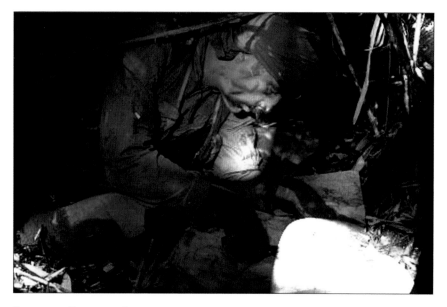

Even a small flashlight for reading a map can give away your position at night.

Light

It is imperative to stress the importance of light sources when carrying out any form of visual tracking. Light and heat-source discipline, though important at all times, is crucial at night. As long as visual observation remains one of the main reconnaissance methods, concealment of light signatures will remain an important camouflage measure. Lights that are not blacked out at night can be observed at great distances. For example, the human eye can detect camp fires (light) from distances up to 8 kilometers and vehicle lights from up to 20 kilometers. Threat surveillance can detect heat from engines, stoves, and heaters from these distances as well. Make sure to only allow smoke and electric light use under cover. When moving at night, vehicles in the forward combat area should use ground guides or black-out lights. Control the use of heat sources and when their use is unavoidable, use terrain masking and other techniques to minimize thermal signatures from fires and stoves.

Author's Note: Many years ago, when fighting in the Oman War, a group of around forty of us set off after dark to reach and establish an airfield at a place known as 'White City' We walked through enemy held country for around four hours before finally reaching the location. A security perimeter was established and most of us set to work rapidly clearing a runway so that at first light a small *'Skyvan'* aircraft could bring in ammunition, water, and supplies. For the first few hours all went well, until, to our horror, we realized that some idiot had decided to make a brew (cup of tea) using an exposed fire. Within minutes mortars started falling and by first light we were being fired at from just about every direction. In fairness, the brave SOAF pilot managed to land the first Skyvan only to get got shot to pieces as it tried to take off. The idiot who had alerted the enemy to our presence by lighting the fire was immediately dismissed from the Army and returned to the UK for discharge. The point here is that the small fire in a valley had been seen some three miles away by the enemy.

NOISE DISCIPLINE

Noise can carry a long distance under certain conditions, humans are good at detecting various types of noise. There is a big difference between normal animal noise and man-made noise. When tracking, any noise should cause a halt to listen and ascertain the direction it originated from. It is not just the noise of the quarry or an enemy you should concentrate on; you need to make sure your own team is not making any noise.

While it may sound a bit silly, one thing I learned early in my military career was noise discipline. That is to say, keep the noise around you to a minimum. In a tracking team, each person is responsible for themselves as well as for the rest of the team members.

Many years ago while in training, we were made to put on a full kit and then jump up and down to see what rattled or made noise. This stuck with me and I still do it before I go hunting. You forget about the small change in your pocket, the spoon you put safely inside your mess kit, or the loose metal catch that keeps hitting your magazine pouch. Jumping up and down will highlight all these small noise pockets. The solution is simple: remove them, replace them, or tape them so they cannot make noise.

SUMMARY

For the professional visual tracker, there is a requirement for skills such as navigation, camouflage and concealment, and movement, as well as light and noise discipline. A soldier tracking a quarry cannot forget that the enemy may lie in ambush or place booby traps in his path. Perhaps the most important of these is camouflage and concealment, although silent movement is also extremely helpful.

While staying on the spoor is your primary goal, the visual tracker should not forget the basics of survival, and that your chances of survival are far greater if you learn how to conceal yourself from your quarry.

Chapter10

ADVANCED TRACKING

As you can see from the previous chapters in this book, tracking people, animals, or vehicles is a wide-ranging subject. While much is given over to explaining human and animal tracking, if the truth be known, visual tracking is a dying art, as technology has entered the arenas.

The reasons for this are fairly simplistic: the quarry we seek today is not that of yesteryear. Today we follow terrorists, criminals, and animals for a wide variety of reasons; some covert, others overt. Terrorism is a world-wide scourge and it is incumbent on governments to keep an eye on those that would possibly do them harm. Those that have committed crimes and may commit more are tagged so that the authorities are aware of their location, while dwindling animal species are tagged in a similar way to help understand their habitat and reverse their declining numbers.

To keep our borders safe and restrict the illegal movement of people from one country to another, we employ vast numbers of agents whose job it is to locate these infiltration points and apprehend the illegal aliens. As small wars ebb and flow around the world, our soldiers sometimes find themselves looking down at the ground for sign of the enemy; unfortunately in the current conflict in Afghanistan this is more to protect themselves from IEDs than to carry out any visual tracking.

By now you will have read the basics of all types of tracking, yet there is still so much more to learn. Skills take years to perfect, but through trial and error of those that have gone before you, leave behind

SAS: In jungle training, make sure that if you run into an ambush you respond quickly.

their expertise. At the end of this book there is a list of available reading material, most written by experts in their own field. I would suggest that if you wish to practice visual tracking, you read as much as you can.

Having written this book, I have tried desperately to do it in such a way as to be a step-by-step guide covering all the basic tracking skills of the British Special Air Service (SAS). Hopefully, you as the reader will understand what 'sign' is before you go looking for it. The book explains how anyone can improve both their ability and awareness in following a quarry through a series of progressive outdoor activities, both in the wilderness and the city.

This book will also appeal to the thousands of armchair warriors who simply dream about living this way of life; it does contain many genuine methods of tracking techniques. In the chapter on advanced tracking I hope simply to depart some of my own knowledge and tips which I have gained over the years.

CHARACTERISTICS OF A VISUAL TRACKER

The visual tracker must be mentally and physically prepared for the hunt. They should be in good physical condition with excellent reserves of stamina, alertness, and to travel for days under hostile conditions. They should also have a very high tolerance to mental stress while maintaining passionate concentration and an eye for detail. Above all, the visual tracker must know what to look for and where to look.

When learning about visual tracking, make sure to take your time and do so in short easy steps. That is to say, get one or more of your friends to make a trail for you, one that is short and with plenty of spoor to follow. The first follows should not be more than 500 meters, with the quarry being no more than thirty minutes in front of you. Build on this so that your quarry can make a track early in the morning which you follow later in the afternoon. Once again, make sure there is lots of clear sign to follow. If the spoor is too weak, then the tracker will become discouraged. Even when the spoor is clear it is easy to miss much of the sign, but don't worry too much about this as long as you manage to complete the track.

Remember, don't just look for sign; try and estimate the direction of travel while anticipating natural openings and exits. Don't look down unless you wish to closely examine a particular sign, keep your head up and look 15 to 20 meters ahead. Check if you hear a noise in front of you or suddenly smell an odor in the air.

Avoid moving from one clear sign to another; spending all your time searching for footprints is not tracking. Tracking means noting what is out of context in nature and realizing the cause. That said, you should always keep a mental note of the last clear key sign you encountered. These will almost always be fairly obvious, such as a footprint in mud or a broken branch, were the quarry tried to secure themselves down a steep bank. Where the quarry is armed, i.e, enemy soldiers or guerrillas, many will use their weapon as a crutch when climbing a steep bank, so check for butt marks. Again, wounded quarry may leave a trail of blood, and this is fairly easy to track; don't just look on the ground, if the quarry is wounded in the upper torso the blood may be on bushes or leaves.

If the blood trail is fairly fresh it is also possible to determine the nature of the wound. Blood that drips steadily is coming from a large open wound, normally in the chest or gut. A wound in the lungs will leave pink bloodstains that are bubbly and frothy. Head wounds produce heavy, wet, and oily blood, while the blood from the abdomen will have a strong odor. With common sense and a refined understanding of the basics you can track most any quarry. Constant practice will develop your skills and lead to success. The following tips below are from the Selous Scouts training. They were possibly the most effective military tracking unit in the world, and this sums up the true characteristics of the visual tracker:

A clear blood trail, which is dark and sticky indicating it possibly came from a head wound.

- Be able to move slowly and quietly, yet steadily, while detecting and interpreting signs.
- Avoid fast movement that may cause you to overlook signs, lose the spoor, or blunder into a terrorist unit that is counter tracking.
- Be persistent and have the skill and desire to continue the mission even though signs are scarce or bad weather or terrain is destroying the spoor.
- Be determined and persistent when trying to find a spoor you have lost.
- Be observant and try to see things that are not obvious at first glance.
- Use your sense of smell and hearing to supplement your sight and intuition.
- Develop a feel for things that do not look right. It may help you regain a lost trail or discover additional spoor.
- Know the terrorist, his habits, equipment, and capability.
- Trust your stick to back you up and your other troops to protect you.

Visual tracking equipment is fairly basic, but using a walking stick with measured marking on it is a great help, as is a smartphone for taking pictures.

VISUAL TRACKING EQUIPMENT

While the professionals will say they just use what they have to track their quarry, if you are starting out or even a good tracker, you will need a few basics. For me these include a modern walking stick, one that I can point with and one that I can use to gently push foliage out of the way without too much disturbance. I have also stuck a cloth measuring tape to the stick, which I can then use for checking both footprint and stride size.

If you do not want to go to the trouble of sticking measuring tape to your walking stick, you will still need some way of measuring the prints and stride, so I would recommend a small cloth tape like those used in ladies sewing kits. The reason for this is noise; an initial pull out auto-return tape measure makes a lot of noise.

Author's Note: Some people will think that having identified a key print, it is now easy to follow. While this may be true, in the case of a military patrol where all members are wearing the same type of manufactured boot, some will be larger than others and a good tracker will be able to identify and confirm this by using a measuring tape:

- A detailed map and compass or GPS unit for navigation would be my second priority.
- Knowing where you are and anticipating the terrain ahead is a major part of tracking.
- A simple writing notebook and pencil is ideal for making quick notes to aid your memory.
- I also like to carry a small flashlight so I can check into dark places under foliage. Be careful when using any form of illumination at night as it may reveal your location to the quarry.

A good tracking kit is very much dependant on the type of terrain over which you will be tracking your quarry. In addition to items that will assist you in tracking, you should also consider taking with you a small survival kit. It is easy to get carried away while concentrating on tracking the quarry and you can find yourself in some odd situations—it is always best to err on the side of providence.

Check your equipment and pay special attention to camouflage and noise. Make sure you are only carrying the essentials and check the rest of the team for the same.

Tracking is not just about following the spoor; you may be entering enemy territory or carrying out cross border operations. In this case you will need to survive as well as track. This means locating drinkable water, and safe places to harbor up. Also check out the local weather for the next few days, as inclement weather can seriously affect any tracking operation.

While tracking a quarry there is a need to know where you are at all times; this is particularly important if you are a military tracking unit. Any tracker that follows a quarry without knowing their rough location is simply asking to get lost—in some environments this could be extremely dangerous. A professional tracker will know (roughly within half a mile or so) exactly where they are at all times and have the ability to withdraw should it become necessary.

SEEING THE ABNORMAL

One thing humans are particularly good at is spotting the abnormal—I put this down to our inherent curious nature. A clown performing in a busy street full of people is easy to recognize; likewise few people would walk past a euro or dollar lying on the sidewalk.

You should adopt the same skills when doing any visual tracking. When looking at the immediate terrain in front of you, always be alert to anything that does not belong or is out of place. We are so used to seeing rubbish everywhere we pay little or no attention to it. When tracking anything that does not belong in the natural environment it should be treated as sign.

Seeing the abnormal is not as difficult as one might think, an unbroken spiders web is a clear indication that no one has passed that way for a while.

HINTS AND TIPS FROM MY TRACKING DAYS

Most of my tracking tuition took place in the jungles of Northern Malaysia. We were lucky to have two excellent instructors from the New Zealand SAS plus several native trackers. For those who have never been in thick jungle, it's fairly difficult to describe. In some places the vegetation is so thick you need to force yourself through it, while in other places the trees are well spaced and resemble a normal forest floor. The terrain is almost always sloping, either sideways, up, or down, making walking difficult. There is only one piece of advice I can offer, and that is *'don't fight it,'* as you will *always* lose.

Personally, I came to love the jungle. It is colorful, interesting, and in some places beautiful beyond words. There are places where man has not set foot, and there are places with many tracks most of which lead to a village or *'ladang'* (area cleared for farming). Additionally, the jungle is cross grained with streams and rivers, some of which are extremely wide and dangerous to cross.

It is an SAS policy that you cross your area of jungle without walking on jungle tracks, man-made or animal. This meant you avoided being ambushed . . . it also meant a lot of hard, tiring work to silently move through the bush. From time to time you could not avoid coming across a jungle track and the procedure was to examine it for prints. If prints were found, then you would move back into the bush and loop back to the track every so often to check on the spoor. Well used tracks are fairly clear and it is easy to see boot prints, bare footprints, or animal sign. I also found it easy to observe things such as spider webs and see if they had been broken. Our native tracker pointed out to me that most of the indigenous women would walk distinctly pigeon-toed—it took me a while but I could eventually tell the difference.

A barefoot native can still leave behind an individual sign.

On one tracking trail, my partner and I became lost and headed for the river from where we could re-orientate ourselves. Even before we had reached the river my nostrils took in a perfumed smell: it turned out to be soap and this successfully guided us to our base camp.

FACTORS WHICH INFLUENCE TRACKING

Without sounding repetitive, there are several factors that will greatly influence your tracking abilities. For example, the size of the quarry—both physical and numbers—will play a major part in the relevant sign left behind. For example, ten people moving fast will leave a lot of sign, but an individual will leave little . . . but if that individual is large, obese, wounded, or moving very quickly, the amount of sign left behind will be increased.

Terrain also has a massive impact on sign. For example, it is easier to track through jungle than a rocky, mountainous area. Where the quarry simply touches a hard surface during their movement there is little to see; where they push through thick undergrowth there will be a much greater amount of sign. While this may seem obvious to the professional visual tracker, the type of terrain will adjust their approach to the whole tracking scenario.

Weather also plays a major part in visual tracking; heavy rain will hide or obliterate sign and make it difficult to see. Bright sunlight has a similar effect upon the human eye and over time, both sun and rain can destroy or erode the spoor. Sunlight will also have some effect on reading spoor. If you are tracking into the sun and are experiencing difficulty in seeing the sign, look back over your shoulder every few yards to confirm your spoor.

Time is also against the visual tracker, as time will repair the natural environment. All sign in the initial stages is placed into a time bracket, this is the time lapse between the earliest possible time that sign could have been made and the time it was located. Once again, this time lapse will play a major role in how the visual tracker proceeds.

THE EFFECTS OF TIME

It is vital that the visual tracker has some understanding of the effects of time on various types of sign as this will help age the spoor. For example:

- Grass blades will remain green for about a day after being trodden down.
  - o Also note when vegetation has not been disturbed, indicating the quarry did NOT go that way. This will help you locate your direction of travel.

- Depending on the amount of moisture in the surrounding earth, footprints in mud will usually take about an hour to fill with water.
- Dew usually stays on for about four hours after sunrise. Disturbed dew drops on grass and plants will indicate passage of something within the last few hours.
- Overturned rocks take a couple of hours to dry in direct sun.
- The dew on cobwebs will last until around midday in bright sunlight, longer if shaded.
- Animal prints overlaid on the spoor will tell you that the spoor was made before nightfall, as most animals move at night. The opposite is also true where a footprint has imprinted over that made by an animal.
- Knowing the time it last rained will also help you estimate the age of the spoor.
- Broken branches and twigs start turning brown roughly ten hours after they were damaged.

Finding a camp can provide so much intelligence on a quarry.

- If you discover a resting area, check the campfire's heat. Place your hand about 5cm (2 inches) over the fire. If the heat is intense enough to make you remove your hand, the fire is less than an hour old; if it feels hot then several hours have passed; if you can touch the embers and they have no heat the fire is more than twenty-four hours old.
- Sudden changes to the wildlife noise and birds taking flight ahead of you will indicate possible human movement.

CAMP INTELLIGENCE

When tracking an enemy force, there is normally some intelligence that has been provided. For example, the size of the enemy force can be estimated by the amount of any fire-fight they may have engaged in. Their withdrawal to a safe location will generally be known; most of all their professionalism and capabilities will have been established.

Armed with this knowledge, the tracking party can proceed to follow the spoor accordingly.

As the tracking party follows the spoor it will—as normal in any tracking—gain additional intelligence. Finding a place where the enemy has rested for any length of time is an ideal place to glean information. The best place is one where the enemy has slept overnight.

With careful observation, the tracker can establish how many soldiers are in the enemy party by counting the places where it is obvious someone has slept, the position where an individual soldier rested the butt of their weapon, the food they ate, where they went to the toilet, who smoked cigarettes, and a lot more. Wounded enemies may have bled during the rest and some may even have died, so make sure to the area for any rapid burial or hidden corpses.

On military tracking operations always look for weapon sign, this one will show the three points of contact with the ground when it is removed.

Camp intelligence will reveal so much about the enemy and help you complete your picture of the size and condition. It should also disclose if they are living off the land or if they are provisioned. Glean every piece of intelligence you can before continuing your track.

Finally, relocate the exit point where the enemy left; this will normally be along the same line of travel as before.

EXHAUSTED QUARRY

Both man and animal will lose some physical control of their body when they become exhausted. In the first instance they will become unstable, that is to say their legs will be weak and they will stagger to some degree. The professional tracker will recognize this in the footfall pattern. The quarry's footprints will cross over and the normal straight line walk will give way to

Both tracker and quarry can become exhausted.

a side-to-side sway, this becomes prominent the more exhausted the quarry is.

Similar disruption to the quarry stride will be found with those that are sick or wounded or in some cases inebriated with drugs or alcohol. Carrying a wounded soldier on a litter will also leave a lot of sign.

SUMMARY

Whenever the opportunity arises, always study the landscape and try to gauge the direction you would take to cut across it.

This book is intended to give the reader not just an insight but also to help learn and understand human, electronic, and animal tracking principles. While the bulk of electronic tracking is designed to be hands off and done remotely, human and animal tracking in the wild requires extra skills. This is especially true in the field of military tracking where the quarry is likely to be armed and dangerous. If you decide to take up visual tracking as a hobby, then here are a few games you can play in your everyday life.

One of the first skills a tracker must obtain is one of good observation. No matter where they are or what they are doing, their mind should be systematically recording places, events, and people with great recall accuracy. Try doing the same when you enter a room for the first time or spot a car number plate and register the number of windows in the house you just passed. The trick is to keep as much information in your brain for as long as possible. Like most subjects, the more you practice the better you become.

If at any time you find yourself looking over a landscape, try to imagine that there is a quarry heading away from you or simply deduce the route you would take if you had to walk across the landscape.

Use logic to understand things. How did they do that? How can I do this? What if I go in totally the opposite direction, not just physically but mentally? Is there another way to solve this situation? Try some *'out of the box'* thinking. Observe people and their habits.

Learn to listen to your sixth sense or analyze any gut feeling you might get. There are many basic traits in the human brain that warn us of danger; learn to recognize them and take appropriate action. You may walk down a street and recognize the same person you saw only an hour ago in a different part of the city—is this coincidence? Take a mental note of their dress, height, and character.

Only ever take calculated risks, never be a gambler. With a calculated risk you can spot the drawbacks and adjust your plans accordingly. Always analyze your actions, basing them on solid information. If you take a gamble, you only need to fail once.

Finally, a tracker has many things to consider, therefore they must be patient, persistent, and have an acute sense of observation. They must possess certain qualities such as an excellent memory, better than average eyesight, fitness, and intelligence; above all they must have a true understanding of nature.